CHASING THE **MOON**

ROBERT STONE
AND **ALAN ANDRES**

CHASING
THE **MOON**

AN EPIC RIVALRY

A MONUMENTAL CHALLENGE

THE RACE TO BE THE FIRST

WILLIAM
COLLINS

William Collins
An imprint of HarperCollins*Publishers*
1 London Bridge Street
London SE1 9GF
WilliamCollinsBooks.com

First published in Great Britain by William Collins in 2019

2020 2022 2021 2019
2 4 6 8 10 9 7 5 3 1

First published in the United States by Ballantine Books, an imprint of Random House,
a division of Penguin Random House, LLC, New York

A catalogue record for this book is
available from the British Library

ISBN 978-0-00-830787-5 (hardback)
ISBN 978-0-00-830786-8 (trade paperback)

Chasing the Moon is available on Blu-ray and DVD
Chasing the Moon is a Robert Stone Productions film for *American Experience*

Book design by Simon M. Sullivan
Typeset in Fairfield
Printed and bound in Great Britain by CPI Group (UK) Ltd, Croydon, CR0 4YY

For my mother, who awakened her ten-year-old son
in the middle of an English midsummer night to
watch Neil Armstrong and Buzz Aldrin make history
as they walked on the Moon.

—R.S.

For Charlie, older brother and teenage rocket
scientist, who, as the youngest accredited journalist
covering the launch of Apollo 14, backpacked and
hitchhiked 1200 miles to Cape Kennedy.

— A.A.

CONTENTS

July 16, 1969

THE SUN BEGAN rising over the northeast coast of Florida on what would be a humid subtropical mid-July morning. The brown pelicans swooping over the dunes sensed the day was anything but typical. Parked tightly together on the sides of the narrow roads to the beaches and causeways were thousands of cars and campers. The air carried the thrumming sound of helicopters ferrying visitors to the Kennedy Space Center. Since shortly after sunrise, all the roads to the Cape had become clogged with traffic.

Nearly a million people were gathering under the harsh Florida sun to witness the departure of the first humans to attempt a landing on another world, the Earth's moon, 239,000 miles away. Should it be successful, the piloted lunar landing would culminate a decade of mounting anticipation.

Eight years earlier, President John Kennedy stood before Congress and called for the United States to put a man on the Moon and return him safely to Earth before the end of the decade. At the time, the Soviet Union's commanding position in space appeared assured. They had been the first to launch an artificial satellite, the first to fly past the Moon, and, in a one-hundred-eight-minute flight, the first to orbit a human around the Earth. When Kennedy addressed Congress, the suborbital flight of the only American to venture into space had been less than fifteen minutes.

But by July 1969, as America readied to launch Apollo 11 from the space center bearing the name of the late president, the Soviet space threat had receded. This would be the twenty-first piloted NASA space mission; in comparison, the Russian total was twelve, and all had remained in low earth orbit. Now the richest nation on Earth was about to undertake a daring technological feat of unprecedented magnitude, a demonstration of national will framed as a world media event. It was a story of courage, adventure, and scientific exploration as well as an exercise in geopolitics.

If not for the persuasive influence of a select group of visionaries, this moment in history would have been inconceivable. For some it was the fulfillment of a personal dream dating back to childhood. For others it was a way to promote democracy, extend human knowledge, and establish the nation's technological and academic preeminence. Some saw it as humanity's inevitable evolutionary destiny; others their personal patriotic duty.

A few hundred feet south of the giant Vehicle Assembly Building, where the Saturn V moon rocket had been put together, a balding middle-aged Englishman dressed in a gray suit looked toward the launchpad from CBS News's temporary broadcast facility. Arthur C. Clarke had imagined this day since his teenage years, having co-authored one of the first serious publications about a possible expedition to the Moon. He was scheduled to appear on live television with CBS's veteran newsman Walter Cronkite shortly after Apollo 11's lift-off. During the preceding year, Clarke's reputation as a leading author of science fiction had risen to greater prominence due to his association as co-writer of director Stanley Kubrick's 2001: A Space Odyssey, the top-grossing film of 1968. For Clarke, the moon landing would establish the extraterrestrial presence of Homo sapiens in the cosmos, with a promise of what the human race might achieve in the future.

At the same moment, in NASA's Launch Control Center, a few hundred feet away from the press center, Wernher von Braun also looked toward the launchpad, three and a half miles away. One of the most famous men in America, the tall, handsome German-born rocket engineer trained his field glasses on the 365-foot-tall Saturn V. Designed by a team under his direction, it was the most powerful rocket in the world. Much like his friend Arthur Clarke, von Braun had his fascination with space travel ignited at an early age by reading science fiction.

A decade earlier the German had been hailed a national hero following the successful launch of the United States's first satellite. A master salesman, von Braun had a unique skill, persuading powerful decision makers—senators, presidents, generals, and dictators—to give him whatever he needed to build his powerful rockets to explore space. However, despite his renown and having lived in the United States for nearly a quarter century, questions surrounding von Braun's past continued to shadow him. During World War II, he had overseen the creation of the first long-range guided ballistic missile for Adolf Hitler's Third Reich. This portion of his biography was well known and had even served as the basis for a Hollywood feature film, but despite having addressed it repeatedly in the past, von Braun and his wartime career in particular remained controversial and open to occasional speculation.

Adjacent to both the designated press area and the Launch Control Center were a few simple wooden bleachers, not unlike those found alongside a typical high school football field. This was NASA's VIP viewing area, the most coveted location on Cape Kennedy, despite its lack of shade on this sweltering July morning. Ambassadors, senators, congressmen, corporate heads, authors, and celebrities had begun assembling there less than an hour before Apollo 11's scheduled liftoff time at 9:32 A.M.

Press photographers had trained their lenses on comedian Jack Benny and the popular host of *The Tonight Show*, Johnny Carson, when their attention was diverted by the arrival of former president Lyndon Johnson and his wife, Lady Bird. This was a rare public appearance for Johnson, one of his first since departing the White House six months earlier. Though the name of President Kennedy would always be linked to the national initiative to place a man on the Moon, Johnson's role as chairman of the National Space Council while vice president and later during his presidency was nearly as decisive.

Johnson climbed the bleachers, stopping to greet old Washington colleagues. Near the top of the stand was Army chief of staff General William Westmoreland, until recently commander of the half million American troops in Vietnam. Directly behind President Johnson and Lady Bird was seated a man far less recognizable to the general public but the one person the former president had relied upon to make this day a reality. Unlike the astronauts or von Braun, former NASA administrator James Webb had never appeared on the cover of a national magazine or been given extensive television exposure. But Webb had been the unwavering choice of both presidents Kennedy and Johnson as the best person to run the American space program during its most challenging and transformative years.

When Americans heard Kennedy's 1961 proposal for a hugely ambitious lunar program, they were uncertain precisely how NASA might accomplish this task and what its final expense might be. The rockets, the spacecraft, and much of the supporting technology to achieve this goal on schedule had to be invented and refined before the end of the decade. No less a space visionary than Clarke and von Braun, James Webb combined insight from his experience running a major federal government agency with years of management experience as a corporate executive. Under his leadership through most of the 1960s, NASA had accomplished a series of spectacular space milestones and had overcome a devastating tragedy that had threatened to end the moon program, only to arrive at this day.

Nearly seven thousand names appeared on NASA's official VIP list; however, far fewer were present in the viewing stands. Notably absent were two prominent Washington politicians. President Kennedy's only surviving brother, Senator Edward Kennedy of Massachusetts, had chosen to view the launch from the congressional offices in Washington. President Richard Nixon was also in Washington, watching the event on television at the White House with Frank Borman, the astronaut who months earlier had commanded the riskiest space mission ever attempted, the first human voyage to the Moon. Nixon had wanted to dine with the astronauts at the Cape the evening before the launch, but the plan had been canceled by the flight surgeon, ostensibly to limit the astronauts' exposure to germs prior to their departure.

At Cape Kennedy and around the world, as the countdown clock reached zero, time seemed to stand still. Millions watching the live television broadcast held their breath and trained their eyes on the picture coming from Pad 39A. Held down by four massive arms at the base of the pad, the Saturn V trembled as its massive engines began to fire, soon building to a combined force of 7.6 million pounds of thrust, at which point the quartet of restraining mechanisms released their grip simultaneously. Three and a half miles away, those assembled began to hear the delayed rumble of the engines as they increased in volume and to feel a steadily building vibration through the soles of their feet.

Slowly lifted on a blinding pillar of fire, the Saturn V began to ascend heavenward as it fought to escape the gravitational confines of the Earth. The people watching carried within their genetic code an inherited biological disposition to dream of a better future and persevere against difficulties and constraints that could seem as daunting as the natural forces pulling against the combined power of the Saturn's engines. The hope and aspirations of countless previous generations had powered the human chronicle to reach this moment invested with so much promise.

CHASING THE **MOON**

CHAPTER ONE

A PLACE BEYOND THE SKY

(1903–1950)

THE BOOK IN the shop window caught the boy's attention immediately. The vibrant purple dust jacket depicted a bullet-shaped machine trailed by an arc of orange flame. If there was any doubt in Archie Clarke's mind what the illustration was intended to depict, the book's title, *The Conquest of Space,* made it obvious. It was a spaceship. The captivating image reminded him of the colorful covers of the American science-fiction magazines that he had seen in the back room of Woolworth's.

Spectacled fourteen-year-old Archie peered into the small W. H. Smith bookshop located a short distance from his grandmother's house on England's Bristol Channel. Dressed in short pants and an Oxford shirt, Archie was walking with his aunt Nellie toward Minehead's main shopping arcade. It was 1932, and Archie had recently lost his father, after a long illness exacerbated by injuries sustained during a German poison-gas attack in World War I. He routinely visited his grandmother and aunt in the small but active southwest England beach resort during the weekends and holidays, leaving his mother freer to attend to his younger siblings on the remote family farm a few miles away.

Nellie Willis, a tall young woman with an intelligent face framed by a brown bob, doted on her clever nephew. She could see that the book displayed in the shop window fascinated him. Despite the family's struggling finances, she gave Archie the "six and seven"—six shillings and seven pence—to purchase it and bring it home.

But when Archie opened the book, he discovered it wasn't the adventure story that he had expected. Instead, he was looking at a book detailing the fundamentals of rocket science—astronautics—supplemented with an imagined account of the first journey to the Moon. Until this moment Archie had assumed space travel was a fantasy. Now he learned that it was actually possible for humans to leave their planet and explore space and that it could happen in the not-too-distant future.

Decades later, as one of the world's leading masters of science fiction and the co-author of *2001: A Space Odyssey*, Arthur C. Clarke would point to that day in 1932 as the moment his life changed. His imagination had been energized by a book, prompting him to wonder about what it might be like when humans began to explore space. In the early 1930s, few in government, media, or business regarded human space travel as a serious possibility. But in Archie Clarke's mind it held transformative and liberating options. If the human species could escape the confines of gravity, was it conceivable that other fantastic possibilities might come to pass in the near future as well?

Archie's fascination with the promise of space travel would motivate and determine the direction of his life following that chance encounter with *The Conquest of Space*. He joined a small cadre of visionaries, theorists, and space-travel advocates whose youthful dreams, curiosity, and determination led directly to humanity's first steps on an alien world only three decades later.

The theoretical mathematics upon which all rocket science was based had begun to circulate in prominent scientific journals only a decade before the publication of *The Conquest of Space*. In the early twentieth century, three independent-minded theorists, intrigued by the idea of space travel after reading works of science fiction as adolescents, attempted to solve the theoretical physics necessary to carry out an actual escape from Earth's gravity. Working autonomously, Russia's Konstantin Tsiolkovsky, American professor Robert Goddard, and physicist Hermann Oberth in Germany conducted their research and study in each of the three countries that would later witness the most decisive events of the early space age. All three theorists were social outsiders intrigued by utopian ideals, and each harbored a personal belief that space travel would inevitably transform human destiny.

The first stirrings of the modern space age arose not in a wealthy industrial nation but in agrarian czarist Russia. At the turn of the century, a popular spiritual philosophy called cosmism—a mixture of elements from Eastern and Western thought, animism, theosophy, and mystical aspects of the Russian Orthodox Church—had influenced a new generation of writers, scientists, and intellectuals. For Russian cosmists, space travel would be the ultimate liberation; once the shackles of the Earth's gravity had been removed and humans inhabited space, the souls of the dead would be resurrected and all humanity would partake in cosmic immortality.

The founding philosopher of Russian cosmism, Nikolai Fedorov, a noted librarian and scholar, chose to personally tutor a bright but impoverished teenager who had been prohibited from attending school due to severe deafness. The student, Konstantin Tsiolkovsky, had an eccentric and strikingly independent intellect and within a few years

Russia's Konstantin Tsiolkovsky, a rural Russian schoolteacher whose 1905 paper first introduced the mathematical equation on which all rocket science is founded. A utopian, Tsiolkovsky believed that human space flight would lead to universal happiness. In a letter he wrote, "Earth is the cradle of humanity, but one cannot live in a cradle forever."

was hired as a small-town schoolmaster, despite his disability. In his spare time, Tsiolkovsky conducted independent research on many scientific subjects, including space travel. He had read Jules Verne's *From the Earth to the Moon* during his adolescence, and later he even tried his hand at writing his own fictional scientific romances.

In 1903 Tsiolkovsky published a scientific paper that contained the first appearance of what came to be known as the "rocket equation," a mathematical formula comparing a rocket's mass ratio to its velocity, the essential calculation necessary to determine how to escape a planet's gravity. Unfortunately, the importance of his publication went unnoticed; the Russian scientific community ignored his work, dismissing it as the musings of an amateur. His paper would remain unread for another twenty years. Undaunted, Tsiolkovsky continued his studies, going on to publish nearly four hundred scientific papers on such matters as space-vehicle weightlessness, the operation of multi-staged launch vehicles, the orbital dynamics of differing rocket burns, and the scientific advantages of polar orbits.

A full decade after Tsiolkovsky's groundbreaking paper, in 1913, a French aircraft designer named Robert Esnault-Pelterie independently published his own version of the rocket equation. But once again few took notice. In the United States, Robert Goddard, the second of the three pioneers of rocketry and a part-time instructor and research fellow at Clark University in Worcester, Massachusetts, was quietly conducting his own rocket research. Entirely unaware of Tsiolkovsky or Esnault-Pelterie, he submitted patent applications for both a liquid-fueled rocket and a multi-stage vehicle.

Like Tsiolkovsky, Goddard had also experienced social isolation during his formative years. A frail only child, he was kept out of school for extended periods due to ill health. As a result, he didn't graduate from high school until age twenty-one. During his solitary time at home he read stacks of books from the local library, particularly volumes from the science and technology shelves. He also read H. G. Wells's science-fiction classic *The War of the Worlds,* which made a lasting impression. At age seventeen in 1899, while aloft in the branches of a cherry tree on his family's New England farm, Goddard experienced an epiphany that moved him so deeply that he noted the date on which it occurred.

Clark University professor Robert Goddard, who in 1926 launched the world's first liquid-fueled rocket from a farm in Auburn, Massachusetts. Throughout most of his career he revealed few details about the progress of his research. However, spies in the United States working at the behest of the Soviet Union and Hitler's Germany attempted to breach Goddard's wall of secrecy.

"As I looked towards the fields at the east, I imagined how wonderful it would be to make some device which had even the possibility of ascending to Mars. I was a different boy when I descended the tree from when I ascended for existence at last seemed very purposive."

During World War I, while teaching at Clark University, Goddard obtained research funding from the War Department for an experimental tube-launched solid-fuel rocket rifle, an early version of what would become the bazooka. He also proposed a rocket that could ascend seventy miles into the atmosphere and carry high explosives or poison gas at least two hundred miles. But after the Armistice, no American military officials thought long-range missiles a subject worthy of further research, so Goddard sought to find support elsewhere.

It was a technical paper funded and published by the Smithsonian Institution in 1920 that suddenly placed Goddard's name on newspaper front pages around the world. "A Method of Reaching Extreme Altitudes" proposed that a rocket could be used for the scientific exploration of the atmosphere, placing artificial satellites into orbit, aiding weather forecasting, and physically hitting the Moon. His paper made no mention of human space travel to the Moon; however, many newspaper reports heralded his study with dramatic headlines implying

Goddard was working on a moon rocket that would transport human passengers.

Within weeks, *The New York Times* announced that a twenty-four-year-old pilot of the New York City Air Police had willingly volunteered to be the first person to fly to Mars. Concerned that the United States must maintain its position with other nations in the air, Captain Claude Collins said he would ride the world's first interplanetary rocket, provided a ten-thousand-dollar life-insurance policy was part of the arrangement. The *Times* treated Goddard somewhat less admiringly than it did Captain Collins when it published a scathing editorial taking the college professor to task for believing that a rocket would function in the vacuum of space. The *Times* slammed Goddard, claiming he was unfamiliar with basic Newtonian physics and showed a "lack of knowledge ladled out daily in high schools." His pride wounded, Goddard soon grew wary of the popular press. Sensational stories about the American professor's forthcoming moon-rocket flight continued to appear in publications around the globe throughout the early 1920s. And occasionally Goddard was complicit, supplying dramatic quotations apparently intended to entice potential investors, such as his plan for a giant passenger rocket capable of crossing the Atlantic Ocean in a few minutes.

However, when Goddard actually made history with the world's first successful launch of a liquid-fueled rocket on March 16, 1926, in Auburn, Massachusetts, no journalists were present, and no account appeared in contemporary newspapers. The date is now celebrated as the dawn of the space age, but for most of his career Goddard carefully guarded information about his research, afraid that others might steal his secrets and profit from his work.

THE SENSATIONAL ATTENTION accorded Goddard's Smithsonian paper appeared in the European press just as the third of the trio of rocketry pioneers, a former medical student from Austro-Hungaria named Hermann Oberth, was readying his work for academic review. Born in 1894, Oberth was a brilliant student of mathematics and had been fascinated by the idea of spaceflight since age twelve, when he com-

Hermann Oberth photographed in his workshop while assisting on the production of the German science-fiction feature film Woman in the Moon. *When he was ten years old, the telephone and the automobile first appeared in his rural hometown. In his later years he witnessed the launches of Apollo 11 and the space shuttle* Challenger.

mitted to memory passages from Jules Verne's *From the Earth to the Moon*. Oberth had tried to interest military strategists in a proposal for long-range missiles during the First World War, but his paper went unread. After the war he revised his work, this time focusing on the basic mathematics underlying space travel. However, when he submitted the paper as his doctoral dissertation, it was rejected as "too fantastic."

Undeterred, Oberth obstinately continued to pursue his studies independently, dismissing his instructors as unworthy to judge his work. For this gifted mathematician, formulating the necessary calculations for space travel was a diverting intellectual exercise that gave him a sense of ownership and agency. "This was nothing but a hobby for me," he said, "like catching butterflies or collecting stamps for other people, with the only difference that I was engaged in rocket development."

He asked himself a series of questions that would need to be answered if humans were to enter outer space: Which propellant should be used—liquid or solid? Is interplanetary travel possible? Can humans adapt to weightlessness? How might humans nourish themselves in space? Can humans wearing space suits venture outside vehicles? In contrast to Goddard's more cautious approach, Oberth embraced the unknown by posing imaginative questions prompted by his reading of science fiction. He then devised practical solutions founded on his mathematical and engineering expertise.

In 1923 he published a short technical version of his dissertation, *Die Rakete zu den Planetenräumen (The Rocket into Interplanetary Space)*, personally paying the expense of the book's entire printing. Fortunately, his vanity-publishing project was a wise decision. By issuing the book in German, which in the early twentieth century was the dominant language of the scientific community, Oberth established himself as the world's leading theorist of human spaceflight, overshadowing the more reclusive Goddard. When he read the German monograph, Goddard believed Oberth had borrowed his ideas without proper attribution, though there is little evidence to support his suspicions.

The Rocket into Interplanetary Space appeared at a moment when many Germans were hungry for something bold, dynamic, and modern to restore the nation's pride, following the defeat in World War I. The 1920s were a time of experimentation in art, film, and architecture and the arrival of new consumer technologies like radio, air travel, and neon lighting. The speed, power, and streamlined design of rockets became associated with a future of exciting possibilites. Less than four years after the publication of Oberth's book, the *Verein für Raumschiffahrt*—or Society for Space Travel—was formed in Germany, and it soon became the world's leading rocketry organization. It published a journal, held conferences, and conducted research experiments. But the stunts of Max Valier, one of the society's founders, were what drew the greatest media attention: He strapped himself into a rocket-powered car and hurtled down a racetrack, trailing a cloud of smoke and flame. Such daredevil exploits proved to be an effective way of generating publicity but did little to boost the society's scientific reputation.

Not long after the society's founding, the noted Austrian-German filmmaker Fritz Lang approached it for technical assistance in connection with his forthcoming science-fiction space epic, *Frau im Mond* (*Woman in the Moon*), a follow-up to his international hit *Metropolis*. Lang hired Oberth, the society's figurehead president, to be the film's technical adviser. The film's studio also engaged Oberth to build a functioning liquid-fuel rocket to promote the movie's premiere, a project that, despite providing the society needed research-and-development money, was unsuccessful.

Albert Einstein and other scientists were among the celebrities who attended the film's opening, but the only rocket to be seen that night was the one that appeared on the screen, created by the studio's special-effects department. Although *Frau im Mond* wasn't a critical hit, it was historically important for introducing the world's first rocket countdown. Fritz Lang created it as a dramatic device to instill suspense in the final moments before the blastoff. It was such an effective way to focus attention and convey the sequence of procedures prior to liftoff that rocket engineers around the world immediately adopted it.

Meanwhile, in the Soviet Union, reports of Goddard's moon rocket and Oberth's scientific monograph prompted Russian space enthusiasts to stake their country's claim by recognizing that Konstantin Tsiolkovsky had been the first to mathematically publish the rocket equation. Tsiolkovsky was in his mid-sixties when he received his vindication, and it coincided with a brief and bizarre moment of space-travel mania. After the First World War, revolution, and a civil war, Russia was in the throes of change, as audacious and provocative new ideas permeated the culture; among them was a renewed interest in utopian Russian cosmism, and a desire to explore new worlds. One of Tsiolkovsky's leading Soviet advocates rallied followers with the slogan "Forward to Mars!"

In 1924, Russian magazines and newspapers reported that Goddard was about to shoot his rocket to the Moon or, in fact, may have already done so. Many Russian readers assumed that colonizing the planets was imminent. At space-advocacy lectures and public programs—including one with a crowd so keyed up that a riot nearly took place—curious attendees demanded to know when flights to the Moon and planets would commence and where to volunteer to be among the first settlers. But after learning that trips to the planets were at least a few decades away, the crowd dispersed in disappointment. In Moscow, an international space exhibition attracted twelve thousand visitors, and a Russian Society for Studies of Interplanetary Travel was founded. But Stalin's rise to power and the beginning of the Five-Year Plan brought an end to Russia's short experimental post-revolutionary sojourn. Despite his new fame at home, Tsiolkovsky received little recognition abroad.

Goddard's reticence for publicity may partly account for the reason that, unlike in Germany and the Soviet Union, no comparable rocketry fad occurred in 1920s America. Instead, a different and more long-lasting phenomenon, which proved influential for the emergence of the space age, arose in the United States: the publication of the first popular science-fiction magazines. In 1926, Hugo Gernsback, an immigrant who had built a business issuing cheaply printed magazines about electronics and radio, introduced *Amazing Stories,* a specialty-fiction publication for which he coined the term "scientifiction." Not long after, Gernsback hired a young technical writer named David Lasser to serve as the editor of a new publication, *Science Wonder Stories.* Lasser, the son of Russian immigrant parents, had enlisted in the Army and experienced combat during the First World War by the age of sixteen. Following months of hospitalization due to the injuries he sustained in a poison-gas attack, Lasser used a disabled-veterans scholarship to attend and graduate from the Massachusetts Institute of Technology. Avid readers noticed that, shortly after Lasser's name appeared on the masthead of the Gernsback magazines, their literary quality improved significantly.

Lasser had become intrigued by press accounts about Goddard, Oberth, and Germany's *Verein für Raumschiffahrt,* and in April 1930 he and fellow New York science-fiction writers and editors formed the American Interplanetary Society. Much like the *Verein für Raumschiffahrt,* its American counterpart aimed to stimulate awareness, enthusiasm, and advocate for private funding of rocket research—while also expanding the readership for Gernsback's magazines. Goddard informed Lasser that he approved of the American Interplanetary Society's mission but abstained from becoming a member. The Clark University professor apparently feared that if it were known he associated with science-fiction enthusiasts, research grant donors might question his judgment and reluctantly withdraw their support.

In one of his first roles as the president of the American Interplanetary Society, Lasser presided over a special event held at New York's American Museum of Natural History: a lecture about space travel, featuring one of the first American screenings of Fritz Lang's *Frau im Mond*. Though a modest-sized audience had been expected, nearly two thousand curious New Yorkers converged on the museum. The only way to accommodate the sizable audience was to add a second screening later that evening.

David Lasser had concluded that there was a need for a book written for curious readers that explained in realistic, accurate, but understandable scientific terms the fundamentals of rocket science, why constructing an operational vehicle should be attempted, and what piloted space travel would mean to humanity. Lasser believed that once humans departed their home planet, a philosophical and political shift would occur throughout the world as people began to perceive the Earth as a small, fragile, isolated sphere in the emptiness of space. This change in thought, he concluded, would lead to the erosion of the dangerous nationalistic and tribal divisions that had brought about the recent carnage of the First World War. He wanted his book to provide the fundamental scientific concepts while forgoing the higher mathematics that might intimidate some readers.

Researching the book, Lasser gathered recently published technical papers from leading scientific journals and corresponded with rocketry activists around the world. He wrote it during the immediate aftermath of the Wall Street Crash, a time when he and many other Americans hoped for a better future. Lasser's optimism colors his imaginary account of the moment when news from the first lunar space travelers is received on Earth: "We learn that wild excitement prevails all over the globe. . . . We cannot but feel now that this journey has served its purpose in the breaking down of racial jealousies." Elsewhere he writes that space travel will result in a new planetary outlook, the realization that "the whole Earth is our home."

Unfortunately, the early years of the Great Depression were not a good time to publish such a book. Lasser and members of the American Interplanetary Society financed the publication of *The Conquest of Space*, but sales were modest. The British rights were sold to a small but venerable publisher, which issued a few thousand copies. Serendipitously, one crucial copy found its way into the hands of teenage Arthur C. Clarke after being displayed in the bookshop window in southwest England.

WHEN HE READ Lasser's book, Archie Clarke was already familiar with the world of American science-fiction magazines. Unsold copies returned from newsstands and drugstores were used as ballast in the holds of the great transatlantic liners sailing between New York and Great Britain. Once they arrived in England, the magazines were sold in shops for a few pence each, including the Woolworth's store across the street from Archie's grammar school, where he often searched through piles of American detective, Western, and romance pulps for the newest science-fiction issues. He soon amassed a substantial collection and compiled a catalog of his reading, scoring stories with a grade ranging from F (fair) to VVG (very, very good).

But when he read *The Conquest of Space* he realized for the first time that "space travel was not merely fiction. One day it could really happen." Shortly before reading Lasser's book, Archie had been fascinated by Olaf Stapledon's *Last and First Men,* an ambitious philosophical novel contemplating the evolutionary fate of the human race hundreds of thousands of years hence. Lasser's suggestion that space exploration would signal the transformation of the human species was a provocative idea, and Clarke yearned to see it happen in his lifetime. He wanted to meet and exchange ideas with others who also shared these dreams of space and adventure.

A small, unelectrified, three-hundred-year-old stone farmhouse in a southwestern English village was the unlikely home where one of the twentieth century's most visionary minds began dreaming about humanity's destiny in the stars. Archie Clarke's parents had both been telegraph operators at different branches of the General Post Office, where, prior to the First World War, they had conducted a covert courtship via Morse code when not under the gaze of their supervisors. Archie had been born while his father, a lieutenant in the Royal Engineers, was stationed in France, later to return badly disabled.

Like many other curious boys, Archie had gone through an early fascination with dinosaurs, an interest sparked at age five when his father casually handed him a cigarette card illustrated with a picture of a stegosaurus. Clarke later attributed his passion for scientific subjects to that moment with his father. An intense interest in electronics, chemistry, and astronomy soon followed, and with the aid of an inexpensive telescope he began mapping the features of the Moon in a composition book. A private grammar school in a neighboring town awarded him a full scholarship, and although he was socially at ease with these more privileged classmates, he was aware he was different. In appearance and background, Archie's modest bucolic home life set him apart. He usually arrived at school wearing unfashionable short pants and large farm boots, which often carried the lingering odor of the barnyard.

Despite his excellent grades, he knew there was little likelihood he could obtain a university education, due to his family's financial circumstances. He loved reading stories in the American magazines that asked "What if?" and subtly questioned conventionally accepted assumptions and rules—both scientific and cultural. In particular, he was immensely impressed by one short story that sympathetically attempted to portray a truly alien "other" and prompted the reader to try to understand distinctly non-human motivations and thought processes. Clarke found within the pages of the science-fiction magazines an invigorating American sense of optimism and intimations of a future with greater opportunities. And before long they also provided a pathway to a community of like minds.

BY THE TIME Clarke read *The Conquest of Space,* almost all publicly sanctioned rocket activity in Germany was nearing an end. Max Valier, world-famous for his rocket-car exploits, was killed during a test of an experimental liquid-fuel rocket engine in 1930—the first human casualty of the space age. A rift had also developed among the *Verein für Raumschiffahrt*'s officers. One faction thought rockets should be used

Members of the German rocket society Verein für Raumschiffahrt. *Hermann Oberth stands to the right of the large experimental rocket, wearing a dark coat, while teenage Wernher von Braun appears behind him, second from the right.*

for scientific exploration, not as weapons, while others urged the society to partner with the German military.

As Europe entered the Great Depression, the society's officers who favored ties with the military exerted greater control and obtained the use of an abandoned German army garrison near Berlin in which to conduct their experiments. Headquartered in an old barracks building, a dedicated corps of unemployed engineers built a launch area and a test stand—a stationary structure on which a liquid-fuel rocket engine could be tested under controlled conditions. All of the serious, highly dedicated engineers were unmarried young men who chose to live with military-like discipline. None either smoked or drank.

Among the most active of the young engineers, one man stood out. An intelligent, blue-eyed, bright-blond-haired eighteen-year-old aristocrat, Wernher von Braun had chosen to dedicate his life to making space travel to other planets a reality. At the old army barracks and rocket testing ground, he acquired valuable hands-on experience designing and launching prototype liquid-fuel rockets in collaboration with Oberth and the other engineers. Von Braun's dedication and ambition soon caught the attention of a group of men who had arrived one morning to observe a test of one of the new rockets. Though dressed as civilians, they were officers from the German Army's ordnance ballistics-and-munitions section, quietly conducting research into future weapons. The Treaty of Versailles, which ended World War I, had imposed restrictions on German military rearmament. However, since rocketry was a new technology not specified in the treaty, rocket-weapons development fell outside its constraints.

In all, the small group of young engineers conducted nearly one hundred rocket launches before the site was finally closed down due to police-imposed safety restrictions and the society's own unpaid bills. But by the time the garrison was shuttered, von Braun had disappeared. Among his former colleagues, it was assumed that he was conducting research elsewhere, though occasional rumors suggested he might be involved in something highly secretive. The extent of the mystery did not become known to the world for another fifteen years.

In debt and its reputation in disarray, the *Verein für Raumschiffahrt* suffered further derision when some members attempted to use the society to promote pseudoscience and nationalist politics. Before his untimely death, Max Valier had endorsed theories about Atlantis, Lemuria, and other popular occultist ideas. One of the society's remaining officers depleted the organization's dwindled funds to finance a public rocket launch intended to prove the validity of the *Hohlweltlehre*—a bizarre doctrine that asserted that the Earth is actually the interior of a giant hollow sphere. Reputable scientists would have nothing to do with it, and the *Verein für Raumschiffahrt* came to an ignominious end just as the new Nazi government imposed prohibitions against any future public discussions about rocket technology or research.

The little news that trickled out of Russia indicated that nearly all interest in rocketry had subsided under the Five-Year Plan. And in the United States, by the time Clarke read *The Conquest of Space,* the life of the book's author had changed dramatically as well. While promoting interest in space travel, David Lasser had discovered he had a nat-

ural talent for organizing people, planning events, and generating
public attention. Not long after the screening at the Museum of Natural History, he began to devote a portion of his time to socialist politics
and organizing to effect political change. Unemployment in the United
States was approaching 25 percent; in Lasser's Greenwich Village
neighborhood, nearly 80 percent of the residents were out of work. He
believed the most important question before the country at that time
wasn't human space travel but reducing unemployment. For the moment, space would have to wait.

Lasser's boss took a dim view of his political activism. Hugo Gernsback wanted him in the office every day, editing the latest issue of *Science Wonder Stories,* rather than taking time off to consult with the
mayor of New York on unemployment issues. Exasperated, Gernsback
summoned Lasser into his office and told him, "If you like working
with the unemployed so much, I suggest you go and join them." Fired
by the world's leading publisher of science fiction, Lasser's short career
as America's first advocate for space travel came to an end as well. His
career change took him to an important job in Washington, D.C.,
where he was tapped to run the Workers Alliance of America, a trade
union for those temporarily employed by the Works Progress Administration of President Franklin Roosevelt's New Deal.

At nearly the same moment that he dismissed David Lasser, Gernsback made a second decision that would significantly impact the life of
Archie Clarke, a continent away. Eager to increase customer loyalty for
his magazines, Gernsback introduced a readers' club, the Science Fiction League, the world's first science-fiction fan organization. Within a
few months it had close to one thousand members, spread among
three continents. Not long after, they began publishing unaffiliated
newsletters, engaging in private correspondence, and traveling to meet
one another.

One of Gernsback's competitors, *Astounding Stories,* started publishing readers' letters in its pages, including the correspondents' ad-
dresses. While poring through one of those issues, Archie Clarke read
that a British Interplanetary Society had been founded in Liverpool a
few months earlier. Now sixteen and having suddenly discovered a

community of like minds, Archie wrote to the British society's secretary, volunteering his services. "I am extremely interested in the whole
subject of interplanetary communications, and have made some experiments with rockets." To impress the society's board, he added that
he had "an extensive knowledge of physics and chemistry and possess
a small laboratory and apparatus with which I can do some experiments."

Within two years he had assumed a position of leadership as one of
the society's most influential board members.

On its final transatlantic voyage, in 1935, the Cunard–White Star
liner *Olympic* arrived in New York City, a day late after encountering
severe February winds. Journalists who met the ship at the pier reported that one of the passengers, Mr. Willy Ley of Berlin, age twenty-
eight, would be spending seven months in the United States, working
with Americans on a project to transport the mail by rocket.

One of the founders of the *Verein für Raumschiffahrt*, Ley had studied astronomy, physics, and zoology at the universities of Berlin and
Konigsberg, and by the mid-1920s he had become one of Germany's
leading advocates for human spaceflight. He was among the society's
strongest voices against rocket-weapons research, believing that rockets should be used for peaceful scientific and exploratory pursuits exclusively. Ley had been disdainful of Max Valier's stunts with his
rocket-powered car but was all for raising public awareness about
spaceflight via popular entertainment. He worked with Fritz Lang during the making of *Frau im Mond* and had become the filmmaker's close
friend. Ley had even written a science-fiction novel, *Die Starfield Company*, an adventure in which the hero battles with space pirates that
also includes an interracial love story and a parable about international
cooperation.

In his leadership role with the *Verein für Raumschiffahrt*, Ley had
been in contact with rocketry researchers and space-travel societies
around the world, and he continued to do so until 1933, when the
Nazis prohibited any exchange of technical and scientific information

about rocketry with citizens of foreign countries. A strong believer in furthering intellectual inquiry through the free exchange of ideas, Ley was deeply bothered by what was happening in Germany.

He watched as scientists and researchers in many disciplines were purged from German academic institutions—primarily for racial reasons—and learned that selected scientific publications were being removed from library shelves. On university campuses, the Nazis conducted public book burnings. Besides scientific works by Albert Einstein and Sigmund Freud and literature by Bertolt Brecht and Thomas Mann, the Nazis had also consigned many classic works of science fiction to the bonfires.

As he read the news and talked with acquaintances, Ley was alarmed as things he had long opposed were gradually accepted as part of everyday life: a cult of loyalty and blind patriotism, militarism, anti-globalism, superstition, and pseudoscience. While Germany touted its reputation for excellence in the sciences, Ley observed how politics had begun to encroach on the scientific method, and positions formerly held by Jewish scientists were filled by less qualified opportunists. His friend Fritz Lang had already fled Germany, and Ley decided he had no other choice but to do the same. He would pretend to leave for a brief vacation in England but knew it likely he would not return home for years.

Members of both the British Interplanetary Society and the American Rocket Society—the new, more serious-sounding name of the American Interplanetary Society—came to Ley's aid by securing him a visa and writing letters of support. Although he arrived in New York with little money, Ley was a recognized expert regarding recent rocket development in Germany and elsewhere in Europe. He was embarking on a new life, believing he would now assume a similar role in the United States.

Ley's initial business venture in the United States generated publicity for collectable rocket-mail postal covers, but unfortunately it returned little income. He next attempted to find employment as a rocketry engineer but was surprised to encounter pervasive skepticism that a rocket could operate in the vacuum of space. Far more welcoming was the small community of science-fiction magazine editors and publishers, and out of necessity Ley began to support himself through his writing.

By the time Ley had arrived in the United States, Robert Goddard had become increasingly reclusive, having moved all his research to a secluded desert testing facility in Roswell, New Mexico. Noted aviation philanthropist Harry Guggenheim had stepped in to provide funding for his rocket research, thanks to the intercession of famed airman Charles Lindbergh as well as one of Goddard's former students at Clark University, an aviation pioneer named Edwin Aldrin. (Aldrin was to become famous a little more than three decades later as the father of astronaut Buzz Aldrin, one of the first two men to land on the Moon.) Goddard's continued secrecy aroused the suspicions of Ley, who considered the professor's reputation in America overrated and unequal to the stature of Oberth. As he began to publish freelance articles about the current state of rocket development for American periodicals, Ley seldom gave Goddard's work equal attention.

No LONGER IGNORED in his homeland, Konstantin Tsiolkovsky was celebrated as a national hero upon his death at age seventy-eight in 1935. One of his last projects was serving as the scientific adviser on a Russian feature film about the first trip to the Moon, *Cosmic Voyage* (*Kosmicheskiy Reys*). Little seen outside the Soviet Union, the adventure film was conceived as socialist-realist entertainment intended to interest young moviegoers in space science. It featured a spaceship named after Stalin, a launch system that used a massive ramp that towered over downtown Moscow, and cinema's first depiction of a flag-raising on the Moon's surface.

Had *Cosmic Voyage* been released in British cinemas, there is little doubt Archie Clarke would have been among the first to buy an admission ticket. Instead, his attention was focused on another film released at nearly the same time as *Cosmic Voyage*. Not long after turning eighteen, Clarke attended a screening of the new British film *Things to Come*. It was a rarity for its time: a serious science-fiction film with a screenplay by a major author, H. G. Wells. *Things to Come* presents a chronicle of the next hundred years, beginning with a devastating second world war that commences on Christmas Day 1940, followed by an extended second dark age and a subsequent technological renaissance in the mid-twenty-first century. In the film's concluding sequence, preparations are under way for the first trip to the Moon. After decades of warfare and barbarity, humanity turns toward outer space to express its innate aspirational yearning. The camera focuses on actor Raymond Massey in the final scene, as he looks heavenward and asks, "All the universe, or nothing. Which shall it be?"

Clarke often spoke of *Things to Come* as his favorite movie of all time. But when the film appeared in English theaters during 1936, audiences would have seen it bookended by newsreels showing labor strikes, militarism in Germany and Japan, and the Italian Army at war in Ethiopia. A glimpse of a technologically advanced future that Clarke yearned for was envisioned on the movie screen, but Wells's screenplay implied that rockets to the Moon would only happen after a devastating world conflagration and a second dark age.

The world was in crisis, but Arthur Clarke sustained his optimistic belief in a better future with a growing library of American science-fiction magazines. His network of science-fiction and rocketry enthusiasts continued to expand, and even Ley became one of his correspondents, not only offering firsthand information about recent rocketry development in Germany but also serving as Clarke's American source for the latest magazines. No longer would he need to haunt the back tables at Woolworth's.

The summer that *Things to Come* was playing in cinemas, Arthur Clarke moved to London to begin his professional life as a junior auditor for the board of education. He had aced the civil-service exam with a perfect math score. "I prided myself on having the fastest slide rule in Whitehall, so I was usually able to do all my work in an hour or so and devote the rest of the day to more important business." The more important business was assuming an active role with the British Interplanetary Society, where he had risen to secretary/treasurer.

On a chilly winter morning a few months after his arrival in London, Clarke and a few friends caught a train out of St. Pancras station to attend a conference in Leeds. The event, held in the city's Theosophi-

Eighteen-year-old Arthur C. Clarke photographed himself using an automatically timed camera shutter in his childhood home in southwest England. The shelves of one bookcase held his extensive collection of American science-fiction magazines.

cal Hall, brought together a handful of young men interested in spaceflight and science fiction for what was later recognized as the world's first scheduled science-fiction convention. The entire attendance was fewer than twenty people. They heard Clarke announce that the British Interplanetary Society planned to move its center of operations from Liverpool to a branch office in London, which shortly thereafter became the society's official headquarters. The new London address was, in fact, a small flat that Clarke shared with the society's publicity director, another aspiring science-fiction writer, William Temple.

But within a few months, the society suffered a major setback. Like their American counterparts, the British society occasionally conducted public demonstration launches of small experimental rockets. While these events proved an effective way to generate publicity, little attention had been given to safety, and during a demonstration in Manchester three spectators were hit by pieces of an aluminum rocket that exploded on the launchpad. Subsequently, all experimental rocket launches in England were subject to prosecution under a nineteenth-century explosives act. The society had to find a different way to capture media attention. Even though they had limited resources, they chose to shoot for the Moon.

It was a purely intellectual exercise but one that no one had attempted before. Working as a team, the core members of the society outlined the many scientific, engineering, and intellectual challenges that a group planning a piloted expedition to the Moon would need to address. They even tried to construct a few working instruments, including an inertial guidance system that would indicate the spaceship's position in space. Assuming they had an unlimited budget, the society's team proceeded to design a launch vehicle with a combined crew cabin and landing craft. The entire budget the society could actually allocate to their research project was roughly one hundred twenty dollars.

Undaunted, the society's team exploited their available resources: youthful enthusiasm, free time, and a smattering of knowledge in a variety of professional disciplines. One member was an expert on turbine engineering; another was a chemist; a third an accountant. There was an interior designer, who envisioned the spacecraft's living quarters. Not one was a full-fledged scientist, but several had some engineering experience. Clarke oversaw the necessary higher math and the astronomical calculations.

Once a week the society's "technical committee" gathered in the evening to dissect details of the proposed two-week lunar mission, with a brief break for fish and chips from the local pub. For their launch system, they decided to use a series of six solid-fuel stages of diminishing size, which were designed to fire in sequence. The committee had ruled out using liquid propellants, having assumed that moving the fuel through a series of mechanical pumps would be nearly impossible in such a massive vehicle.

When the project was completed, the results were published in the January 1939 issue of their newsletter, the *Journal of the British Interplanetary Society*. The entire print run for that issue filled two cardboard boxes, which Clarke retrieved from the printer and walked back to his flat. But their modest journal generated publicity that reverberated around the world. Initally Clarke and other society members were interviewed by London newspapers and on BBC radio. Next, the *Journal* received attention in the prestigious science magazine *Nature*, which summarily dismissed the moon ship as pure fantasy. The scientific community thought it necessary to silence these starry-eyed young troublemakers before someone took them seriously.

Undismayed, Clarke and his companions returned every instance of public criticism with pointed and sarcastic rebuttals—whenever the publications deigned to give them space to reply. The criticism from the scientific establishment inspired the creation of the first of Clarke's Three Laws: "When a distinguished but elderly scientist states that something is possible, he is almost certainly right. When he states that something is impossible, he is very probably wrong."

News about the society's rocket ship spread internationally. In the United States, *Time* magazine reported on the controversy, and English-language newspapers as far away as India included it in their world-news summary. The society's journal noted with pride that one account "stole half the photo-news page of a national Sunday newspaper from Herr Hitler." During the flurry of publicity surrounding their moon rocket, Clarke and Bill Temple met one foreign-language journalist who made an enduring impression. Early in the interview, Temple began to wonder whether the tall quiet-voiced German might be a Nazi spy, especially when he showed particular interest in their collection of clippings about rockets as weapons. Clarke and Temple agreed that in this instance it was probably wise to avoid impressing their visitor with their knowledge of astronautics. Instead, they pretended to be merely a couple of harmless science-fiction fanboys.

The best-informed members in both the American and British rocket societies continued to assume that all rocket-related research and development in Germany had come to an abrupt end following the rise of the Nazis. Living in the United States, Willy Ley had heard nothing from his homeland to make him believe otherwise. The Third Reich appeared more concerned with rearming its land army and rebuilding its air force than with funding scientific rocket research, which few believed had any practical application as a weapon of war. Ley logically assumed that transporting a small explosive payload via a rocket would be a waste of money, and he was certain that other military strategists would agree. Meanwhile, he hoped he might eventually find a full-time position with an American company interested in developing rockets for scientific purposes. He continued to advocate for space travel, writing articles on a variety of scientific subjects for popular magazines in the hope that an informed public in the United States would avoid being seduced by the pseudoscientific and mystical fads that had become popular in Germany recently.

While on a trip to Los Angeles, Ley was delighted to reestablish contact with *Frau im Mond*'s creator. Fritz Lang's sudden departure from Germany had come shortly after Hitler's propaganda minister, Joseph Goebbels, banned his latest film, in which Lang had put the words of the Nazis in the mouth of an evil criminal mastermind. Lang was now working for MGM, where he had directed his first American film, *Fury*, starring Spencer Tracy and Sylvia Sidney. It was a provocative thriller that addressed the scourge of lynchings in the United States, though told through the eyes of an innocently accused white man. In it, Lang depicted American vigilante mob justice with visual comparisons to what he had witnessed in Nazi Germany.

Sitting on a veranda under a starry California sky, Lang and Ley discussed the impending war in Europe and mused about travel to the Moon and the planets. However, if they had wanted to revisit their earlier cinematic collaboration, finding a copy of *Frau im Mond* would have been impossible. Hitler's Gestapo had confiscated every exhibition print a few years earlier. The film had disappeared.

Not long after the Third Reich's invasion of Poland in September 1939, Britain entered the war against Germany, forcing Clarke and Bill Temple to vacate their Bloomsbury flat and shut down the British Interplanetary Society's headquarters. Should the British forces in Western Europe fail to prevent France from falling, Germany's Luftwaffe bombers were expected to appear in the skies over the heart of London within days. Londoners with the opportunity to do so sought out alternative lodging with friends and relatives in the city's less vulnerable outskirts or moved to the countryside. When Clarke and Temple locked their door, they left behind Clarke's almost-complete run of American science-fiction magazines, a collection numbering in the hundreds that had taken him nearly a decade to assemble. He would never see them again.

The worst of the Blitz didn't come to London until the fall of 1940, when the city was bombed continuously for nearly two months. Arthur Clarke saw none of it; now working for the Ministry of Food, he had been relocated to a seaside resort in North Wales. Sometime in the early spring of the next year, their Bloomsbury flat took a direct hit, destroying everything except the outside walls.

CLARKE SPENT THE early months of the war processing paperwork that documented the precise location of each ton of imported British tea. His position in the civil service gave him a temporary deferment from military conscription, but by the end of the year, service in one of the armed forces was unavoidable. He joined the RAF in the hope that he might be able to acquire a valuable education in the fundamentals of celestial navigation, but instead he was assigned to a technical unit devoted to a new utilization of radar to assist aircraft during poor-visibility landings. It was Clarke's first opportunity to collaborate with another group of trained scientists, a team from the Massachusetts Institute of Technology that had worked on the invention's development.

Corporal Clarke was then assigned to an RAF training center in Wiltshire, not far from Stonehenge, where he taught night classes on the fundamentals of radar. However, the subject of the corporal's classroom lectures frequently turned to astronautics, prompting his students to nickname him "Spaceship Clarke." During a lecture a student might mischievously ask the instructor how a rocket functions in space, setting off a long discussion about multi-stage rockets and reaching the Moon, complete with diagrams and basic calculations. During his off hours he wrote technical articles for journals such as

Electronic Engineering. His career as a published science-fiction author was yet to come, though just prior to joining the RAF he had completed the preliminary draft of his first novel, *Against the Fall of Night.*

As the Allied forces closed in on Germany in late 1944, Clarke and a group of the most active members of the British Interplanetary Society met in a London restaurant one evening. Val Cleaver, a society officer who worked in British aviation, told the diners details about his recent business trip to the United States. While visiting New York, Cleaver had met with Willy Ley and discussed recent reports of a large German rocket weapon that was said to have hit targets in Antwerp and London. Ley had heard reports that it was a frightening and more sophisticated successor to the V-1, a low-flying cruise missile that had appeared in the skies of southern England that summer, sometimes arriving in waves of more than one hundred missiles a day. Ley dismissed the jet-powered V-1 as a crude and inaccurate weapon of little military value, assuring his British guest that the reports of a bigger, high-altitude rocket bomb were nothing more than desperate Nazi propaganda. Cleaver, who had already seen classified U.K. military-intelligence reports detailing the existence of the big rocket, cautioned his friend, "If I were you, I wouldn't be quite so sure."

Laughter was heard around the dinner table after Cleaver recalled his words of caution. But no sooner had the amusement subsided than the gathering was interrupted by the sound of a huge crash outside the restaurant. "The building shook slightly," Clarke recalled. "We heard that curious, unmistakable rumble of an explosion climbing backwards up the sky, from an object that had arrived faster than the sound of its own passage." The abrupt intrusion had been the British Interplanetary Society's introduction to the deadly V-2 rocket, the world's first operational ballistic missile.

Should Ley have needed any further persuasion about Germany's new rocket weapon, a copy of *Life* magazine published a few weeks later would have been sufficient. A double-page spread provided a detailed and fairly accurate cutaway diagram of the V-2 and a graphic illustration presenting its trajectory from launch to impact. *Life* also reproduced military photographs that pictured recovered rocket en-

A German V-2 rocket containing a small explosive warhead is readied for launch during the final months of World War II. More than three thousand V-2s were fired against Allied targets in England and Belgium, but as a strategic military weapon of destruction it was largely ineffective.

gines. It described the V-2 as a "spectacular weapon" but judged it "a military flop." Despite its impressive engineering, the new weapon was an ineffective boondoggle. As Ley had predicted, the V-2's destructive power was limited by its small payload capacity. In fact, fewer military and civilian casualties resulted from V-2 attacks than the total number of slave laborers killed due to the harsh conditions surrounding the weapons' assembly. But decisively, when the German high command chose to fund the V-2 by diverting funding earmarked for fighter-jet aircraft, they ceded the airspace to Allied bombers, thus hastening their own defeat.

Ley published an article about the V-2 in an American magazine just as Allied forces entered Germany. In it he speculated that the large new rockets were the work of Hermann Oberth and thought it unlikely that either Oberth or his associates would survive to tell the story of the V-2's birth. "Those who knew the full story are dead already," he stated. "Those that are still alive will die before the war is over." But far more important to Ley was its legacy: The V-2 had provided undeniable proof that it was possible to launch a large, fully operational guided missile.

Parts of a V-2 confiscated by the Allies were shipped to the United States, where Robert Goddard examined them at the Naval Experiment Station in Annapolis, Maryland. Goddard found the design of the V-2's gyroscopically controlled stabilizing vanes, its fuel-injecting turbopumps, and its combustion chamber remarkably similar to features he had used on the rockets he developed and launched in Roswell, New Mexico. In the mid-1930s, when Goddard had been conducting his research far from the eyes of the press and curiosity seekers, both Hitler's military intelligence organization—the *Abwehr*—and Soviet espionage officials had dispatched spies to gather information about Goddard's progress. But despite Goddard's suspicions that the V-2's design had been stolen from his work, the technology for both rockets evolved along independent parallel tracks, with the Germans already ahead of Goddard by the early 1930s. A few days after Goddard scrutinized the confiscated V-2, Germany fell to the Allies and the war in Europe ended. Already ill with cancer, Goddard would die at age sixty-two four months later. His death came on the same week that the United States dropped two atomic bombs on Japan, ending the war in the Pacific.

In the wake of the German surrender, the United States's joint chiefs of staff immediately approved an unprecedented new program intended to achieve a strategic military advantage over future adversaries by obtaining proprietary access to the Third Reich's advanced weapons technology. Not only were physical weapons and plans to be seized, but the United States's wartime intelligence agency, the Office of Strategic Services (OSS), sought to find the brainpower behind them as well.

The plan progressed so rapidly that the first group of German scientists and engineers arrived on American soil before President Truman became aware of the program's existence. It began as Operation Overcast, an initiative focused on taking possession of Nazi scientific knowledge and technology for use in the war against Japan. However, after the Japanese surrender, the larger program was renamed Operation Paperclip and included many more former Third Reich engineers, technicians, and scientists. The code name arose from the Office of Strategic Services' use of paperclips to mark the intelligence files of scientists and engineers selected for inclusion in the program.

Willy Ley assumed his unique knowledge of rocket science and his experience working with Hermann Oberth would help him obtain a financially secure job with either the United States government or an American corporation expanding into rocket development. But in the eyes of the American military, Ley was an outsider. He learned from contacts in the U.S. government that many of the German engineers who had designed the V-2 had survived the war and had been brought to the United States to work with the War Department. It was cruelly ironic. Ley had left Germany out of conscience, while those who had chosen to remain and build rockets for Hitler were accorded special attention and employed by the U.S. government. Many of Ley's associates from the *Verein für Raumschiffahrt* who had worked on the V-2 would be among those leading the effort to make human space travel a reality. But Ley would not be among them.

From an American military officer, Ley learned that the Nazis' director of the V-2 program had not been Hermann Oberth, as he had assumed. Its manager was Wernher von Braun, who as a bright eighteen-year-old aristocrat and part-time student had been personally introduced to Oberth and the *Verein für Raumschiffahrt* in 1930 by Ley. When recalling von Braun's persuasion skills, Ley wrote his friend science-fiction author Robert Heinlein, "I only hope that the U.S. Army will not suddenly find him 'charming' in addition to being useful."

In the waning days of the Third Reich, von Braun and his top associates had considered their options. Soviet forces were approaching from the east and the American Army from the west; their capture was inevitable. They knew their unique technical knowledge would give them leverage when negotiating terms of surrender. When von Braun polled his group, the consensus was to surrender to the Americans, and after hiding for a few days in a remote area of the Bavarian Alps, they made furtive contact with a U.S. infantry division. By the time the Soviet Army arrived at von Braun's rocket development and testing area at Peenemünde on the Baltic Sea, nearly every one of the top scientists and engineers had already surrendered to the American forces.

Von Braun and more than one hundred other members of his German rocket-development team arrived quietly in the United States a few months after the end of the war in Europe. For decades, significant details about how they and other German scientists were vetted and cleared for entry were shrouded in secrecy. But it is undeniable that the United States government concealed the fact that it gave preferential treatment to some German scientists and engineers who had been Nazi Party members or suspected of complicity in war crimes.

The first public news of Operation Paperclip came in an understated press release issued by the War Department on October 1, 1945. It announced that a carefully selected number of "outstanding German scientists" would be brought to the United States to impart technical knowledge vital to the nation's security. The one-page release said that they would be in the United States on a temporary basis and all had made the journey voluntarily. Not long after, *The New York Times* revealed the "entire German staff at the [Peenemünde] rocket-weapon base, about ninety men," had arrived in the United States. In actuality, during the war as many as twelve thousand had been employed at Peenemünde, but only the top echelon—around one hundred fifty rocket scientists and engineers—had traveled to the United States to work with von Braun.

THE WAR DEPARTMENT and the Office of Strategic Services considered the German scientists and engineers such valuable assets that it was deemed far more important that the United States government gain access to their expertise and knowledge than worry about the controversial—and highly classified—details contained in their wartime files. Stalin's encroachment into Eastern Europe had already prompted fears of a protracted conflict with the Soviet Union. And during the immediate post-war years, Americans suspected of having communist sympathies were deemed a far greater threat to the nation than someone with a past association with the defeated Third Reich.

A few years earlier, David Lasser's tenure as the president of the Workers Alliance of America had ended when members of the Communist Party asserted domination over its leadership. Lasser was a socialist but opposed communism, and he chose to resign from the alliance in protest. President Franklin Roosevelt subsequently asked him to form an organization that would train the unemployed so that they could transition into the workforce. But his nomination ran into trouble when reactionary members of Congress discovered Lasser's name on a list of suspected leftists. While sitting in the gallery above the United States House of Representatives, Lasser listened as a Texas congressman with a reputation for grandstanding and publicly exposing the identities of political subversives attacked his reputation. He ridiculed the author of *The Conquest of Space* on the House floor as "a crackpot with mental delusions that we can travel to the Moon!" The House exploded in laughter and Lasser's nomination died in the midst of the uproar.

Shortly after V-E Day, it appeared that Lasser's fortunes in Washington might be improving, as memory of his ridicule in the House began to fade. The Truman administration asked him to assist with the rebuilding of Europe under the Marshall Plan, offering him a position as a consultant to the secretary of commerce. Ironically, at nearly the same moment that American military and intelligence officers were quietly obscuring the past histories of former Nazi Party engineers, David Lasser's political opponents began circulating false rumors about his alleged past association with subversive political organizations, in an effort to tarnish his reputation. They questioned his loyalty and argued that his "contrary views" posed a serious security risk.

Lasser was incredulous at the coordinated smear campaign. "I kept asking myself, what kind of government would do these things? What kind of people were we that this sort of thing happened?" Despite vigorous support from prominent politicians, the accusations and rumors

The Conquest of Space *author David Lasser (center) and a fellow labor organizer meet with first lady Eleanor Roosevelt in Washington. Not long afterward, Lasser was ridiculed on the floor of the House of Representatives as "a crackpot with mental delusions that we can travel to the Moon!"*

effectively blacklisted Lasser from any further government employment. Far less renowned than the Hollywood Ten or writers like Howard Fast or Arthur Miller, David Lasser, one of the country's first space advocates and the author of *The Conquest of Space*, became one of the first victims of the Red Scare.

The War Department's decision to bring scientists and engineers from Hitler's Third Reich to work for the U.S. government did not go unopposed. Prominent physicists such as Albert Einstein and Hans Bethe as well as former first lady Eleanor Roosevelt criticized Operation Paperclip. But the larger looming reality of the Soviet Union's brutal domination of Eastern Europe, legitimate fears of domestic espionage, and reports of a possible Russian nuclear-weapons program silenced most public resistance to the program. No congressmen delivered speeches questioning whether the German scientists posed a security risk or held contrary political views. Instead, the White House asked the Department of Commerce to issue reports that would explain to ordinary Americans how their daily lives would benefit from

wondrous German technological breakthroughs in food preparation
and the manufacture of cheaper, stronger clothing, such as run-free
nylons and unlimited yeast production.

DESPITE THEIR FRIGHTENING close encounter with the V-2 in London,
Arthur Clarke and the other directors of the British Interplanetary Society were optimistically anticipating the coming rocket age. In particular, they wondered if an increased interest in rockets and space
might affect their post-war careers or lead to entrepreneurial opportunities. Clarke's mind turned in this direction when contemplating pos-
siblities for radio and television communication stations in space.

Early in 1945 he wrote a letter to the magazine *Wireless World,* in
which he proposed a novel idea. If three geosynchronous satellites
were placed in stationary orbits above fixed points on the globe, each
would act like a radio mast erected 22,300 miles above the Earth. Signals sent from a ground station could be received by the satellite in
orbit and then amplified and retransmitted over a third of the globe.
His letter persuasively argued that a technology ostensibly developed
for war could also have peaceful applications far more beneficial to
humanity.

During that summer he expanded the idea into a four-page article,
"The Future of World Communications," which, after being cleared by
RAF censors and re-titled "Extra-Terrestrial Relays: Can Rocket Stations Give World-Wide Radio Coverage?", was published in the October issue of *Wireless World*. Its publication was the first to outline a
geosynchronous communications-satellite network and is now considered one of the landmark technical publications of the twentieth century.

Clarke later jokingly noted that his article was met with monumental indifference and earned him a total income of fifteen pounds. But,
in fact, it was read in the right places. Copies circulated in offices of
the United States Navy and within a newly created private nonprofit
called Project RAND, an American think tank designed to coordinate
military planning with research and development.

A second, less historically important technical article written by

Clarke had far greater immediate personal impact on its author. Shortly before Clarke was demobilized in 1946, "The Rocket and the Future of Warfare" was published in *The Royal Air Force Quarterly.* He sent a copy to a young Labour MP, who upon reading it said he wanted to meet its author. Coincidentally, their meeting occurred just after Clarke had been deemed ineligible for a university educational grant. In the course of their conversation at the House of Commons, twenty-eight-year-old Clarke told the MP about his predicament. "In a very short time, my grant was approved and I applied for admission to King's College, London." Rockets had not taken Clarke into outer space, but they indirectly propelled the farm boy toward a university education.

In addition to their technical publications and books of nonfiction, Tsiolkovsky, Oberth, and Ley had written works of science fiction to popularize their ideas of space travel. A similar impetus prompted Clarke to begin work on a second novel. Between semesters, Clarke set aside his studies in physics and math to write *Prelude to Space.* It wouldn't be published for another five years, but it was his first attempt to articulate his optimistic vision of the coming space age.

The most commercially successful American work of space advocacy published during the late 1940s was an oversized book written by Willy Ley and illustrated by artist Chesley Bonestell titled, in tribute to David Lasser, *The Conquest of Space.* While its objective paralleled that of Lasser's book published nearly two decades earlier, the Viking Press volume found a much larger readership curious to learn the fundamentals of rocket science and the promise of the coming space age. Bonestell's scientifically accurate astronomical paintings were already familiar to readers of *Life* magazine and *Collier's,* another mass circulation weekly. His work was also known to American moviegoers, though his efforts in Hollywood remained largely unheralded at the time: He had created the architectural renderings of Xanadu in *Citizen Kane* and the futuristic skyscrapers in the film adaptation of Ayn Rand's *The Fountainhead.*

For many children growing up in the early 1950s, the imaginary yet scientifically accurate images in *The Conquest of Space* served as their visual introduction to spaceflight. The book's success led to public-speaking engagements for Ley, including appearances on the emerging medium of television, where he explained what the recent talk about space travel could mean for the future. In his role as a popular science writer, authority on space, and debunker of pseudoscientific fads and occultist beliefs, Ley served as a voice of avuncular reason amid a flood of sensational UFO reports that appeared frequently in newspapers and magazines during the early Cold War era.

Across the Atlantic, producers at the BBC had similar programming needs. When they wanted someone who could clearly communicate scientific ideas to the general public, Arthur Clarke was the person they repeatedly called upon, and he soon established himself as a minor national TV personality. Now living in North London within walking distance of the BBC's television broadcast studio, Clarke appeared not only in his role as the spokesperson for the British Interplanetary Society but also as a frequent guest whenever a producer required someone on short notice to speak about astronomy, space science, physics, or even the fourth dimension. These early appear-

By 1950, Arthur C. Clarke was a frequent guest on the BBC, explaining to British audiences the probable reason for the sudden increase in reports of UFO sightings, how humans might travel to other planets, and differing theories of a fourth dimension.

ances occurred at roughly the same moment as the publication of Clarke's first nonfiction book, *Interplanetary Flight,* a short volume advocating for space exploration.

In Rahway, New Jersey, the son of a garment worker from the Ukraine read about Clarke's new book in an ad published in the latest issue of *Astounding Science Fiction.* The sixteen-year-old was fascinated by news articles about flying-saucer sightings and became intrigued by the possibility of life on other planets. But he knew little of the fundamentals of rocket science or planetary astronomy, so he ordered a copy of *Interplanetary Flight* via mail order from an address in the magazine.

Two decades after Clarke discovered *The Conquest of Space* in the bookstore window, it was his book that fell into the hands of another impressionable teenager. The high school senior, Carl Sagan, would later speak of reading *Interplanetary Flight* as the "turning point in my scientific development," the moment that solidified the course of his life, leading him to become not only a noted astrophysicist but the most recognized popularizer of science in the United States during the last quarter of the twentieth century.

CHAPTER TWO

THE MAN WHO SOLD THE MOON
(1952–1960)

O N A THURSDAY evening in March 1952, viewers of NBC's *Camel News Caravan* were introduced to a man who, in the next few years, would be celebrated as a national hero for ushering America into the space age, becoming his adopted country's most widely recognized man of science. That only a decade earlier Wernher von Braun had overseen Adolf Hitler's most ambitious weapons program is among the strangest and most confounding ironies of twentieth-century history.

It's no surprise that von Braun's affiliation with the Third Reich was not mentioned on the evening of his national TV debut. The handsome forty-year-old wore a tailored double-breasted suit and might have been mistaken for a crusading district attorney in a Hollywood film noir. But in no screen thriller did a DA ever speak in such a distinctly Teutonic accent or display the fantastic props that von Braun held onscreen. Viewers were told that these were models of space vehicles that would transport humans into the cosmos within a few years and bring an end to threats from Iron Curtain nations around the globe. Von Braun was on TV to launch a nationwide publicity campaign for the mass-circulation magazine *Collier's* and, in particular, its latest issue, with a cover that proclaimed, MAN WILL CONQUER SPACE SOON!

In the spring of 1952, television sets were a fixture in one out of every three American households, an increase of 200 percent in the past twenty-four months. The new medium's first users were predominantly upper-middle-class families living near cities with network-affiliate stations. For most of these viewers, von Braun's spaceships were not a new sight. Adventure series such as *Tom Corbett, Space Cadet* and *Captain Video* were already competing against small-screen Westerns to capture the imaginations of young audiences. But von Braun's NBC appearance that evening was intended for their parents, many from the generation of recent war veterans who were redefining America as it assumed its position as a global economic and military superpower. Indeed, the editorial introducing the new issue of *Collier's* delivered an urgent Cold War warning: If the United States did not immediately establish its dominance in space, it would lose this high ground to the Soviet Union. Not only was America's destiny in outer space but the nation's security depended on mastering the science and technology to get us there.

Collier's readers were introduced to von Braun as the technical director of the U.S. Army's Ordnance Guided Missile Development Group. "At forty, he is considered the foremost rocket engineer in the world today. He was brought to this country from Germany by the U.S. government in 1945." Further details about his wartime work were carefully omitted. He was pictured at the head of a table next to his first tutor and mentor in rocket research, Willy Ley. Shortly after the Peenemünde team arrived in the United States, Ley had cautioned friends to be wary of von Braun's seductive charisma. By the early 1950s, von Braun's charm, as well as his considerable political savvy and innate talent to inspire, had worked magic on his former enemies.

This wasn't the first meeting on American soil of the two former colleagues. It was on a December evening in 1946 that Ley and von Braun had looked each other in the face for the first time in more than a decade and a half. Their post-war experiences in their adopted country had differed dramatically. Ley was the son of a traveling salesman; von Braun had been born into privilege, an aristocrat whose father was a politician, jurist, and bank official. Von Braun grew up with a sense of entitlement, which, when combined with his innate charisma, effortlessly opened doors. Physically, he could have been mistaken for a matinee idol; Ley once described von Braun's appearance as "a perfect example of the type labeled 'Aryan Nordic' by the Nazis." In affect and appearance, Ley, on the other hand, personified the "absentminded professor" stereotype. He wore thick-lens eyeglasses and spoke with a heavy accent, which peppered a discussion about UFOs with references to "flyink zauzers." Nevertheless, Ley was a talented communicator with an ability to convey his curiosity and fascination about scientific subjects to audiences, which found his passions infectious. Unfortunately, he was less successful when attempting to find rocketry-related work in the United States in spite of his expertise, while von Braun charmed his way into new opportunities.

Their reunion had occurred when von Braun made his first visit to New York City, to attend an American Rocket Society conference, accompanied by his entourage of military minders. The presence of von Braun's escort didn't prevent Ley from extending an invitation to dinner at his apartment in Queens. Over glasses of wine, the two men talked until nearly 3:00 A.M., catching up on fifteen years of history during a discussion Ley later described as both tense and informative. Von Braun revealed the history of the Nazi rocket program and how he had come to lead it. However, he was less forthcoming about some crucial details that became more widely known only decades later.

Von Braun disclosed the circumstances behind his abrupt disappearance from the *Verein für Raumschiffahrt* in the fall of 1932. A captain in the German Army's weapons department, Walter Dornberger, had personally recruited von Braun to research the development of liquid-fuel rockets as ballistic weapons. Dornberger set up a small lab at Kummersdorf, a secluded estate south of Berlin, and gave von Braun a stipend, a stationary rocket-engine testing stand, and an assistant. They imposed total secrecy on von Braun's work since all Army-funded research was classified. While at Kummersdorf, the young scientist— then age twenty—was allowed to pursue his doctoral studies in physics and engineering at the University of Berlin. It was while he was at work on his dissertation that the Nazis removed Jewish professors and academics with suspected leftist political leanings, and burned books at the public rallies. Von Braun admitted to Ley that he had focused exclusively on his studies and was oblivious to the political significance of what was happening around him. When von Braun's dissertation was finished, the German Army demanded it be titled "About Combustion Tests," in an attempt to disguise the fact that it included detailed information about his liquid-fuel-rocket research at Kummersdorf.

Von Braun explained to Ley how by 1937 the German Army had financed the development of the world's most powerful rocket of that time, a towering twenty-one-foot liquid-fuel missile, which they secretly launched from a remote island on the Baltic Sea. During Germany's period of rearmament, von Braun said, he also worked on developing rocket-assisted airplane takeoffs for the air force, the Luftwaffe. Not long after, a competition ensued between the different branches of the German armed forces, with the Luftwaffe offering von Braun five million marks to establish a new facility for rocket development, and the Army coming up with six million more. The Army's additional one million marks ensured that Dornberger would continue as von Braun's superior and would exert greater control over the combined eleven-million-mark Luftwaffe-Army project. Von Braun couldn't believe his good fortune. "We hit the big time!" he said, and was then tasked with finding the perfect location for his new facility.

Von Braun continued his story as he told Ley how he had searched for a remote secure location near a large body of water. Sensing the coming war, he also thought it should be a site strategically situated for future rocket launches against the Allies. His mother suggested Peenemünde, a relatively uninhabited pine-covered island on the Baltic, where his father used to go duck hunting. The Luftwaffe funded the luxurious facilities, and by 1938 the island had a brand-new town, a chemical-manufacturing facility, a power plant, and its own railway. At full capacity it would house twelve thousand employees.

As their conversation continued into the night, von Braun went on to vividly describe the first test of the A-4 rocket in October 1942. Listening attentively, Willy Ley attempted to mentally record as much information as he could. He had published *Rockets: The Future of Travel Beyond the Stratosphere* only two years earlier and realized the history section of his book was now unacceptably obsolete. Von Braun said the A-4—the rocket that became better known by its propaganda name as the V-2—had been designed to carry a one-ton warhead two hundred miles. Von Braun revealed that the first A-4 had been decorated with a painted insignia that depicted a long-legged nude woman with a rocket sitting upon the crescent moon, a reference to Fritz Lang's *Frau im Mond*. The first test of the A-4 had gone far better than any of his team had expected. It reached an altitude of almost sixty miles. At a celebration after the launch, von Braun said Colonel Dornberger had looked around the room and remarked, "Do you realize that today the spaceship was born?"

After the A-4's successful first flight, a number of guidance-system design problems remained unresolved and months of work lay ahead before the weapon could be deployed in the war. Some weeks later, von Braun and Dornberger were summoned to meet with Hitler to explain the production delays. They showed him a film of the first successful test. When the screening ended and the lights went up, Hitler displayed a sudden new enthusiasm for the A-4 program and talked of it as the superweapon he had been hoping for. Hitler immediately approved further research funding and conferred a professorship on von Braun—"the youngest professor in Germany"—and promoted Dornberger to major general.

The British had become aware of the activity at Peenemünde, however, and on the night of August 17–18, 1943, nearly six hundred RAF bombers dropped hundreds of tons of explosives on the facility. Almost seven hundred people were killed in the raid, most of them foreign prisoners who had been forced to work on the assembly of the early rockets. As a result of the raid, production for both the V-2 and the less complicated V-1 cruise missile was relocated to a distant underground facility. Development and testing of the V-2 missile continued at Peenemünde, but with a smaller workforce.

A crucial part of the story remained untold, however. The new hidden underground production facility was, in fact, built as part of the Dora-Mittelbau concentration camp in the Harz Mountains, more than three hundred miles southwest of Peenemünde. During the final two years of the war, thousands of slave laborers from the Soviet Union, Poland, and France were worked to death at Dora-Mittelbau while building thousands of rockets. A few grim details about the rocket-making facility appeared in American newspapers around V-E Day, but otherwise the full story detailing the extent of the horrors surrounding the V-2's production went unreported in the United States for decades.

Wernher von Braun photographed in the early 1950s holding a model of the V-2 rocket.

Despite holding a high position in the Third Reich, von Braun had been in serious danger. He told Ley how he had run afoul of Heinrich Himmler's SS, which had been competing with the Army for control of the rocket program. After frankly admitting to Himmler that he preferred working under General Dornberger, von Braun was arrested by the SS. For two weeks he was held under suspicion of being a defeatist, a communist sympathizer, and a potential defector. His file even contained a report that during a private conversation he had confessed that if given a choice he would prefer to design spaceships instead of weapons, a comment that was considered dangerously anti-militarist. His release came only after General Dornberger made a personal appeal to Hitler's minister of armaments and war production, Albert Speer, who, in turn, conveyed it to Hitler.

As Allied forces moved toward Berlin and the first V-2s began hitting targets in London and Antwerp in late 1944, slave laborers in Dora-

Mittelbau's massive tunnel facilities were assembling as many as six hundred V-2 rockets a month. But by March of the following year, the Russian Army was approaching Peenemünde, and von Braun and five hundred engineers and scientists fled south to the Bavarian mountains. Hiding in an alpine hotel, von Braun and Dornberger plotted their surrender to the Allies.

Two days before von Braun told his story to Ley, his picture had appeared in *The New York Times* in an article about the Operation Paperclip scientists. The *Times* reported that the technical knowledge of these "former pets of Hitler" would save American taxpayers an estimated 750 million dollars in research-and-development costs. Someone unimpressed by America's new German brain trust was Ley's friend, science-fiction author Robert Heinlein, who was disgusted when he learned that Ley had been "fraternizing with a Nazi." Heinlein wrote to a mutual friend in the Navy that by spending the evening with von Braun, Ley had displayed careless expediency. As a result, Heinlein decided to withdraw his support for Ley's efforts to find a government job.

Culturally, the new global superpower that had welcomed von Braun and the Operation Paperclip engineers still suffered from a pervasive inferiority complex. The superiority of the European tradition in the arts and sciences went largely unquestioned. To the citizens of a nation still less than two hundred years old, Wernher von Braun personified a cultured, well-mannered, soft-spoken European aristocrat, much like the characters played by Cary Grant, George Sanders, Claude Rains, or Paul Henreid in Hollywood movies. Typical of his upbringing, Von Braun was an accomplished musician who could play the "Moonlight Sonata" from memory. He was perfectly cast for the role he was to play in post-war America.

In Washington, America's Joint Intelligence Objectives Agency carefully sanitized the troublesome personal histories of von Braun and other Operation Paperclip engineers, much as Hollywood publicists fictionalized the biographies of actors under the studio system. Von Braun arrived in the United States aware that his wartime management experience guaranteed him a position of importance—a position that until very recently had allowed him to be indifferent to the struggles of others or the ethical repercussions of his personal actions. Unlike most of his fellow space visionaries, von Braun was driven less by a personal desire to make a better tomorrow than by a personal ambition to accomplish something that no one had done previously.

Early in his career, von Braun realized that to achieve his goals, he had to become a persuasive salesman. He learned how to convince the key decision makers that his vision would confer to those in power precisely what he had deduced they most desired. To generals and dictators, he offered a promise of military superiority and national prestige; to those worried about threats from outside enemies, he promised security; to those searching for a sense of purpose and meaning, he promised a unique adventure and the fulfillment of our human destiny. Along with his persuasive salesmanship, he cultivated a rare talent to inspire others to do their best and to instill in them a sense of loyalty and dedication that seldom wavered.

It was while quartered with the other German rocket engineers at the Army's Fort Bliss in El Paso, Texas, that von Braun made a personal choice to become an evangelical Christian. His decision followed a visit to a modest white-framed church situated on a parched Texas lot, where, he later said, he came to realize for the first time that religion wasn't something inherited like an heirloom but a personal commitment requiring effort and discipline. Von Braun's conversion may have served to compartmentalize his European past from what lay ahead, as did his decision, at nearly the same time, to wed his eighteen-year-old cousin and bring her to Texas as part of his new life.

THE U.S. ARMY had shipped hundreds of crates containing the components for scores of confiscated German V-2s to the United States. Von Braun and his team restored their mechanisms and successfully launched them from a test range at White Sands, New Mexico, sending some as high as one hundred miles above the Earth. Later flights tried out a two-stage launch vehicle, with a second, smaller research rocket positioned on the nose of the modified V-2. After climbing to a height of twenty miles, the smaller rocket separated from the V-2 and, using its own engine, achieved a velocity greater than five times the speed of sound and ascending nearly two hundred fifty miles above the Earth.

But within the offices of the Pentagon, there was little interest in large rockets like the V-2, as either an offensive or defensive weapon. Its performance during World War II had proven the V-2 more effective as a weapon of psychological terror than destructive power. Von Braun's work at White Sands had yielded interesting scientific information, but how it might be applied to Defense Department concerns was unclear.

The Cold War had become the dominant concern of those overseeing America's defense planning, and space research had no role in it. Nevertheless, von Braun tried to think of ways in which he might persuade military decision makers to fund his space-flight research and development. He sought out the physicists at the national laboratory in Los Alamos, New Mexico, who had developed the atomic bomb, with a proposal to marry an atomic warhead with one of his ballistic missiles: the genesis of the intercontinental ballistic missile (ICBM). But in the 1940s it was still assumed that conventional bomber aircraft were the most practical and effective way to deliver a heavy nuclear weapon to a target. Von Braun also outlined plans for a large orbiting military space station, which he argued could serve as a bombing platform capable of targeting any location on the globe and a unique surveillance outpost. It was an idea that he continued to refine and lobby for throughout the 1950s.

But his third and most ambitious idea to stimulate space funding didn't depend on government defense strategies at all. Recalling how reading science fiction had fired his youthful imagination, von Braun decided to engage a new generation of space dreamers by writing a novel about the first voyage to Mars. Unfortunately for von Braun, the publishers who read the manuscript found his dialogue wooden and faulted the lack of a romantic subplot. When writing his manuscript, von Braun had emphasized the story's technical accuracy; entertaining his reader was of secondary concern. In all, eighteen American publishing houses rejected it.

Four years after Arthur Clarke and the other officers of the British Interplanetary Society narrowly avoided being killed by the V-2 explosion, Clarke thought it time to exploit the experience to their advantage. He sent a letter to von Braun, offering him an honorary society membership. Von Braun graciously accepted, replying, "Despite the grief the work of me and my associates brought to the British people, [your invitation] is the most encouraging proof that the noble enthusiasm in the future of rocketry is stronger than national sentiments." An exchange soon followed, in which von Braun sent Clarke some scientific details about the recent White Sands tests and Clarke invited von Braun to deliver a paper at an upcoming British Interplanetary Society conference in London.

In the immediate post-war years, von Braun's U.S. Army minders kept him on a short leash. It was a time of heightened fear about Soviet spies operating within the United States, and the Army considered him a valuable asset. During his brief return to Germany to get married, the Army had von Braun under constant watch to prevent a Soviet kidnap attempt. But the Army had other worries as well. Von Braun gave a well-received speech to the El Paso Rotary Club in January 1947, but not long after, reports appeared in newspapers that revealed that some Operation Paperclip engineers had to be sent back to Germany after troublesome details about their Nazi past had come to light. Most press accounts stressed the Germans' eagerness to work for the United States—their anti-communist sympathies were often cited— and indicated their hope to become American citizens. Nevertheless, the same month that von Braun addressed the El Paso Rotary, the president of the American and World Federations for Polish Jews said, "It is a sad reflection and insult to the consciousness of humanity [to welcome] these evil representatives of Nazi science . . . to this country with open arms." For the next two years, von Braun maintained a modest public profile.

After the Germans had concluded their work with the refurbished V-2s at White Sands, the Army had few new projects to keep them occupied. There was scant military funding for additional rocket research, and their quarters at Fort Bliss were needed to house the Cold War's growing roster of soldiers. The Army had to find a new permanent home for its restless and underutilized rocket specialists. In 1950, at the urging of Senator John Sparkman of Alabama, the Department of Defense moved the Army's Rocket Branch of the Ordnance Department's Research and Development Division to the recently shuttered Redstone Arsenal in Huntsville, Alabama. It was here, near the Tennessee River, that the Fort Bliss rocket men relocated to buildings constructed a decade earlier for the manufacture and storage of chemical weapons and munitions. New signs announced the facility as the Army's Ordnance Guided Missile Center. The thirty-five-thousand-acre site on which Redstone Arsenal had been built had already witnessed a great deal of history.

The fertile soil on the southern dip of the Tennessee River Valley had been home to Creek, Cherokee, and Chickasaw tribes prior to the arrival of the Euro-Americans, who forcibly removed all the native peoples during the 1830s and 1940s. For a few years before the Civil War, slaves worked the land's large cotton plantations; however, during Reconstruction the land was subdivided into small tenant farms, many cultivated by the families of the recently freed. But after seven decades, the tenant farmers were forced to relocate when the Huntsville Arsenal and the Redstone Ordnance Plant were built on the land during World War II.

The German engineers who arrived in 1950 found the green landscape surrounding the Tennessee River a welcome change. After sandy and dry El Paso, Huntsville was somewhat reminiscent of Silesia, von Braun thought. When the engineers arrived, Huntsville's population was only sixteen thousand, reflecting a post-war decline following the closing of the chemical-weapons facility.

Huntsville's flagging economy began to rebound once the Army's new Ordnance Guided Missile Center was established at Redstone Arsenal. The city's new citizens brought a bit of European culture to northern Alabama, and local grocery stores began selling sauerkraut. Huntsville took on the air of a New England college town, albeit with a Dixie flavor: It founded a symphony orchestra, a ballet, and a Broadway Theater League and opened a newly expanded public library.

ONE OF THE most popular books in the new Huntsville library's collection was Ley and Bonestell's *The Conquest of Space*. In New York, the

Hayden Planetarium created a popular show based on the book, which subsequently traveled to other cities. As a result of this collaboration, Willy Ley had become friendly with the planetarium's chairman. In the course of a lunch conversation, Ley asked his friend why it was that the British Interplanetary Society could schedule annual conferences about human spaceflight but no such event had ever been planned in the United States.

Without much further discussion the Hayden's chairman simply responded: "Willy, go ahead; the planetarium is yours."

That seemingly minor exchange set in motion a sequence of events that would alter American attitudes toward space travel during the coming decade and turn von Braun into a celebrity of the early television era.

Less than six months after the Hayden Planetarium's chairman gave his consent, Ley had assembled a roster of speakers for the First Annual Symposium on Space Travel, held, symbolically, on Columbus Day 1951. He conceived it as an event that would generate media interest and public awareness. Invitations were sent to every print and TV outlet with an office in the New York City area, including foreign publications. Among the two hundred attendees who heard talks on space medicine, space law, and upper-atmosphere science were two journalists from *Collier's* magazine.

Collier's assistant editor Cornelius Ryan was unimpressed with the report he received about the Hayden Planetarium conference from the two staffers who had attended. The Irish-born former war correspondent had little patience for all the recent talk about human space travel, believing the subject was more appropriate for children's television than for a serious magazine. However, at the insistence of the managing editor, Ryan reluctantly attended a conference on space medicine in San Antonio. *Collier's* also dispatched *Conquest of Space* artist Chesley Bonestell to sit in as well. After the first full day of presentations, Ryan was left confused and unimpressed.

Over cocktails, Ryan began a conversation with a tall handsome man also attending the conference. Grasping his highball glass, Ryan confessed, "They've sent me down here to find out what serious scientists think about the possibilities of flight into outer space." As he gestured around the room he admitted, "I don't know what all these people are talking about. All I could find out so far is that a lot of people get up to the rostrum and cover a blackboard with mysterious signs!" He said *Collier's* was considering publishing a major cover story about space exploration, but he doubted readers would find anything presented at this conference of much interest.

His companion introduced himself as Wernher von Braun, and as he attempted to help Ryan understand the day's presentations, he motioned for two others to join them. One was Fred Whipple, chairman of the Harvard University astronomy department, and the other was Joseph Kaplan, a scientist specializing in the study of the upper atmosphere. Over a lengthy dinner that lasted until nearly midnight, von Braun, Whipple, and Kaplan passionately took turns explaining why they believed humanity's destiny lay in space.

The latest recipient of the von Braun charm offensive returned to New York a true believer. He convinced the magazine's managing editor that a unique *Collier's*-branded space symposium would generate publicity for the magazine and attract advertising dollars away from the emerging threat of television. Ryan insisted that von Braun should serve as *Collier's* key expert, with additional articles written by other specialists from the New York and San Antonio conferences.

At that moment, von Braun and his engineers were spearheading the creation of the Redstone rocket, the Army's first short-range ballistic missile. The Redstone was a bigger but less streamlined variation on the V-2, designed to carry a payload of nearly seven thousand pounds. Its rapid development was part of a newly unfolding rocket rivalry between two different branches of the armed services. At nearly the same time that the Army decided to develop the Redstone, consultants for the U.S. Air Force began work on their own intercontinental-ballistic-missile development program, which would eventually reach fruition with the Atlas. Though designed to deliver munitions, both the Redstone and Atlas would become far better known to the general public a decade later for their role as the vehicles that transported the first Americans into space.

Having brought von Braun's entire team to Huntsville, the Army was now more comfortable with him entering the public spotlight. He had not spoken at either of the two American space conferences during the autumn of 1951, though other Operation Paperclip Germans—Dr. Hubertus Strughold and Heinz Haber—had delivered papers. Strughold had risen to prominence as a leading researcher on the physical and psychological effects of human spaceflight, but details about his past had been deliberately obscured. Indeed, it would be another four decades before allegations of his complicity in notorious Nazi-era human medical experiments were widely published. By 1951, public objection to government employment of the Paperclip scientists and engineers had largely subsided, though the Germans' hopes for American citizenship would remain unfulfilled for a few more years.

In April 1951, journalist Daniel Lang published an extended profile of von Braun in *The New Yorker*. A former war correspondent who had covered World War II in Italy, France, and North Africa, Lang was intrigued by the ethical choices faced by men of science during the Cold War. Lang described von Braun's personality as "exuberant rather than reflective" and thought he comported himself like "a man accustomed to being regarded as indispensable." Unlike the reticence he displayed in later interviews, von Braun was unguarded with Lang, even confessing, "Working in a dictatorship can have its advantages, if the regime is behind you. . . . We used to have thousands of Russian prisoners of war working for us at Peenemünde." The profile also revealed that 80 percent of the Germans working at Redstone Arsenal had been members of the Nazi Party or affiliated organizations and that von Braun had been a party member. Pressed by Lang to address the morality of his decision to work for the German Army, von Braun explained, "We felt no moral scruples about the possible future abuse of our brain child. . . . Someone else would have done the job if I hadn't." After the *New Yorker* profile appeared, von Braun became more cautious when talking to the press.

At the end of the year, *Collier's* gathered a symposium of space experts in the magazine's New York office, where von Braun was joined by Ley, Whipple, Bonestell, and others. In conjunction with the magazine's forthcoming issue, *Collier's* commissioned a series of detailed full-color illustrations of giant spacecraft reaching orbit and showing how humans would live and work in space. These were based on plans and sketches drawn up by von Braun and his team in Huntsville. Von Braun also assigned Gerd de Beek, a former Peenemünde staffer now at Redstone, to construct scale models of the huge rocket and space station. It was de Beek who had painted the *Frau im Mond* insignia on the first successfully launched A-4 in 1942.

In the weeks before the March publication date, the *Collier's* staff concluded that von Braun was such an asset that they chose him as their spokesperson to promote the special issue at media events. After his national television debut on the *Camel News Caravan,* von Braun appeared several more times on the new medium within twenty-four hours. In Manhattan he traveled from one network broadcast studio to another for scheduled live interviews on the most popular programs. In the morning he was on the *Today* show on NBC; by the afternoon he was in a CBS studio chatting with entertainer Garry Moore. He even put in an appearance at the end of the day during a broadcast of ABC's children's adventure series *Tom Corbett, Space Cadet.* Whenever he appeared, he brought along de Beek's models to illustrate his argument.

After working nearly seven years with little national recognition, von Braun was exuberant that his American moment had arrived. "I'm tickled to death about this TV and radio business," he wrote Ryan. "Space rockets are hitting the big time!"

The *Collier's* issue looked like nothing that had been seen previously in a mass-circulation general-interest magazine. Bonestell's color illustrations provided eye-catching visions of space vehicles heading into orbit and space stations under construction. Equally impressive were the magazine's cutaway diagrams showing the interiors of von Braun's huge rocket and the rotating space station. The attention to detail in these images conveyed a convincing sense of accuracy, even though they were based on plans that were entirely imaginary. With a circulation of three million copies and an estimated reach of twelve to fifteen million readers, that single issue of *Collier's* was believed to have been seen by 8 to 10 percent of the American public.

The *Collier's* publicity campaign included department-store window displays and posters on buses, subways, and newsstands. The cumulative effect not only promoted the magazine, it suddenly transformed von Braun into the world's most visible public proponent for space exploration. *Collier's* introductory editorial emphasized space as a national defense and security concern, though military implications were downplayed in the articles. In von Braun's feature article, "Crossing the Last Frontier," he made reference to his orbiting "atomic bomb carrier" space station, but apocalyptic scenarios were kept to a minimum. Not that *Collier's* was averse to exploiting Cold War fears to sell issues; two years previously they had commissioned Bonestell to paint disturbing scenes of Manhattan consumed by an atomic mushroom cloud. But with their space issue, *Collier's* presented a more optimistic technological vision of the future, which promised both adventure and a new domain for human exploration. It was a welcome relief from concerns about nuclear proliferation, Soviet espionage, and the ongoing Korean stalemate.

So favorable and resounding was the reader and media response to that *Collier's* immediately began planning additional space-themed issues. The next one hit newsstands October 1952 and featured articles about the first human voyage to the Moon. Von Braun continued to make public appearances, such as an extended segment on the only network TV program about science, *The Johns Hopkins Science Review,* which broadcast three separate thirty-minute episodes discussing the *Collier's* series.

Culturally, the *Collier's* issues had prompted a significant shift in public attitudes toward human spaceflight: The subject was no longer ridiculed or approached with embarrassment. At the San Antonio conference the previous year, the organizers had been reluctant to use the word "spaceflight" in the event's title, for fear that it would diminish its seriousness. After spring 1952, politicians speaking about space travel seldom encountered the derision David Lasser had experienced on the floor of the House of Representatives a decade earlier.

ON THE EVENING of July 9, 1952, CBS newsman Walter Cronkite sat in a wood-paneled studio in Chicago, covering the opening of the Republican National Convention. It was a decisive moment for the young journalist as well as for the country. Cronkite was anchoring the first live network broadcast of an American political convention. The GOP campaign had begun in New Hampshire the same week that von Braun made his TV debut. Now the field of candidates had narrowed to General Dwight D. Eisenhower, Ohio senator Robert Taft, and California governor Earl Warren.

The former speaker of the House, Joe Martin, Jr., gaveled the convention into session, and in a speech watched by millions, Martin attacked the Democrats in power as "disciples of a dead-end economy" who offered the youth of America nothing more than a "road to nowhere." Martin then segued to offer a more optimistic vision of the future, and when doing so he appeared to have in mind a recent national magazine. "We have an entire new world about to unfold!" Martin promised. "I listen to the words of scientists and engineers. . . . They are optimistic! They are visionary beyond our fondest dreams! They say that science and technical skill are uncovering new horizons that all but defy the imagination." He described new advances in medicine, power, and transportation. Then Martin turned to the heavens. "Travel in space! I mean interplanetary travel—in our solar system—no longer is the figment of a cartoonist's imagination. It is on the verge of reality! Who knows what wonders lie beyond the limit of our atmosphere, what new worlds will open to us?"

A foreigner who had arrived in the United States on the *Queen Mary* a few days earlier watched the convention with great interest on the television set in his hotel room. Arthur C. Clarke was in America to promote the publication of his newest book of nonfiction, *The Exploration of Space*. The judges of the Book-of-the-Month Club had chosen it as a featured selection in the wake of the Hayden Planetarium conference and the enthusiastic response to the *Collier's* issue.

Clarke sat up and listened to Joe Martin with astonishment. A politician was on national TV giving a speech that contained passages that could have been written by Clarke himself. "I didn't expect the Republican Party to take official cognizance of the topics the science-fiction writers love to mull over," he recalled later. He listened as Martin talked of smart "electronic computing machines" and wrist radios utilizing "a new invention no bigger than a postage stamp, called the transistor." He was delighted. "After that, nothing could drag me away from the convention."

This was only the second trip Clarke had taken outside of the United Kingdom, and as he traveled the United States, others aspects of 1950s America caught his attention. At the invitation of Ian Macaulay, a fanzine editor whom he had met at a science-fiction convention, Clarke traveled to Atlanta, Georgia. At that moment Macaulay was active in the fight for civil rights, organizing to end segregation in schools and public transportation and working to increase African American voter registration. Macaulay's activism prompted some lengthy late-night discussions during which Clarke learned more about America's long history of racial inequity.

Prior to coming to the United States, Clarke had never met any black people. He stayed with Macaulay again in Atlanta the following year, and during his return visit he was at work on the final pages of his novel *Childhood's End*. In the book's conclusion, an adventurous scientist named Jan Rodricks is selected by representatives of an advanced extraterrestrial civilization to witness the transformation of the human race into a higher form of life: a collective cosmic entity. Clarke's decision to make Rodricks—his fictional representative of all mankind and "the last person on Earth"—an astronomer of black African descent was a bold and politically provocative choice for 1953.

Since reading a short story about a sympathetic alien in a 1934 issue of *Wonder Stories,* Clarke had been fascinated by science fiction's potential to evoke empathy for alien characters and convey to readers the viewpoint and values of those from other cultures. During its formative decades, the genre's predominantly male readership was hungry for escape from the mainstream culture and curious about emerging technologies and new ideas. Many science-fiction readers were intelligent yet socially marginalized in some way due to a variety of reasons, such as ethnicity, sexuality, religion, or race. Plus, conventional society's habit of ostracizing and calling those with an interest in technical or intellectual subjects "nerds" or "eggheads" was familiar to a sizable segment of the genre's readership. Not surprisingly, therefore, science-fiction readers often identified with the alien.

By the early 1950s, some science-fiction authors had begun using mass-market paperbacks as a literary vehicle to subtly question conventional attitudes about race and sexuality. The cover illustration on a 1952 book of short stories by Robert Heinlein featured the first visual depiction of a black astronaut, even though no such character was specifically described within the book, and publishers' sales directors at the time expressly asked illustrators not to include African Americans on book covers, since it was assumed this would hinder sales south of the Mason–Dixon Line. A year later *Weird Fantasy,* a science-fiction comic book, published an allegory on American segregation in a story about an emerging civilization on a planet where orange robots are accorded privilege over the less entitled blue robots. In the story's kicker ending, the space-suited emissary, who sits in judgment of the planet and denies its application for admission to a galactic republic, is revealed to be a black astronaut.

Robert Heinlein, who had also challenged social preconceptions about masculinity and femininity, wrote *Tunnel in the Sky* in 1955. In the young-adult novel, the race of the central protagonist is not overtly defined, but there are hints that he is black. Heinlein confessed that he used this literary device in an attempt to disarm his white readers, hoping that in the course of the narrative they would gradually come to realize the protagonist's race after feeling empathy and identifying with him.

Writing in an essay for *The New York Times* shortly after his visit to Atlanta, Clarke addressed this issue as part of a larger defense of science fiction as literature. "Interplanetary xenophobia [in earlier science fiction] has given place to the idea that alien forms of life would have as much right to their points of view as we have. Such an attitude . . . can obviously help spread the idea of tolerance here on Earth (where heaven knows it's needed)."

THE SUMMER THAT Arthur Clarke returned to the United States and was finishing the manuscript of *Childhood's End,* he finally met Wernher von Braun, when both were guests at the Washington, D.C., home of the American Rocket Society's president. In a lengthy diversion from a dinner conversation about humanity's future in space, Clarke described his love of scuba diving and explained that it was an effective way to simulate the experience of being weightless in space. He then vigorously urged von Braun to take up the sport for the same reason, which von Braun did only a few weeks later, remaining an active diver for most of his life.

Even while overseeing the development of the Redstone rocket and working on a top-secret plan to quickly and inexpensively launch the first satellite into orbit, code-named Project Orbiter, von Braun continued his public advocacy for human spaceflight. An unexpected opportunity arose directly as a result of the *Collier's* publications, just as the magazine released its eighth and final space-themed issue, featuring a description of the first human voyage to Mars. In early 1954, Hollywood came calling, in the person of Walt Disney, who was interested in producing a series of well-financed hour-long documentary films based on the magazine series. Disney was in the midst of creating a new prime-time TV program, *Disneyland,* which would mix recycled older content and new programming in order to promote another new venture, his California theme park, scheduled to open in 1955.

At the suggestion of one of his top animators, Ward Kimball, Disney approved three space-related "Tomorrowland" episodes adapted from *Collier's* articles. The first episode, "Man in Space," had Ley, von Braun, and Heinz Haber discussing rocketry history, orbital science, and the physical challenges facing humans during spaceflight, before concluding with an animated look into the near future as humans first entered space. Disney's choice to feature three onscreen experts with distinctive German accents became an issue of concern within the studio prior to filming, until it was deemed that their authenticity was more valuable than any possible negative associations. The history portion of the program included World War II–era footage of V-2 launches, yet there was no mention of Nazi Germany's part in rocket development; the V-2 was merely referred to reverently as "the forerunner of spaceships to come." Among the featured experts, it was von Braun who commanded the viewer's attention. Looking into the camera, he confidently asserted, "If we were to start today on an organized and

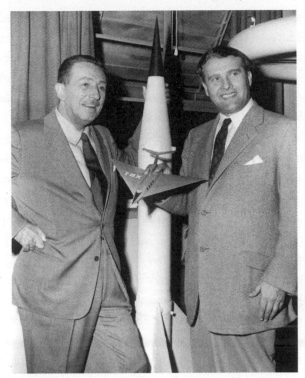

Walt Disney and Wernher von Braun in 1955 collaborating on "Man in Space," one of three space exploration–themed episodes broadcast on Disney's weekly television series in the 1950s. The programs were seen in more than a third of American households, including the White House, where President Dwight Eisenhower tuned in.

well-supported space program, I believe a practical passenger rocket can be built and tested within ten years."

"Man in Space" premiered in the spring of 1955. Forty million households—more than a third of the American viewing public—watched the broadcast on their black-and-white televisions. Polling conducted that year revealed that nearly 40 percent of Americans believed "men in rockets will be able to reach the Moon" before the end of the century, a figure that had more than doubled in six years. Once again, von Braun's space advocacy had political repercussions. One viewer who saw "Man in Space" lived at 1600 Pennsylvania Avenue, and after the broadcast Disney received a request from the Eisenhower

White House for the loan of an exhibition print of the program so that it could be screened for Pentagon officials.

In nearly all of his entertainment, Disney promoted commonly accepted traditional American values. Though he kept his personal political attitudes out of the spotlight, Disney's were conservative and anti-communist. Von Braun's association with Disney therefore subtly bestowed an imprimatur of American respectability on the former official of the Third Reich. And in a final act of assimilation, a month after "Man in Space" aired, von Braun and more than one hundred other Germans working at the Redstone Arsenal in Huntsville appeared in a newsreel taking an oath of allegiance as they became U.S. citizens.

Von Braun was also celebrated in American public schools, many of which had recently introduced new audio-visual equipment into the classrooms. The Disney studio actively licensed 16mm exhibition prints of "Man in Space" as an entertaining teaching aid. Disney published a teacher's study guide to use in conjunction with the screening, which featured von Braun's photo on the cover and suggested classroom discussion questions such as "How likely is it that the present barriers between nations will tend to break down as contacts with other planets develop?"

Von Braun's onscreen presence in the Disney programs coincided with his appearance in another far less visible film. Released in late 1955, the low-budget black-and-white Department of Defense film *Challenge of Outer Space* recorded an "officers' conference" in which von Braun addressed a classroom of officers from various branches of the armed services. The content of his presentation paralleled the Disney films—and even included *Collier's* illustrations. However, von Braun approached it almost entirely in terms of Cold War military superiority, something that was never mentioned in the "Man in Space" program. When discussing his space station, von Braun described it as something of "terrific military importance both as a reconnaissance station and as a bombing platform" with "unprecedented accuracy."

Seated in front of an American flag, von Braun answered rehearsed questions from the officers, such as "Dr. von Braun, can you explain why it would be easier to bomb New York from a satellite than from a

plane or a land-launched guided missile?" His response to such sobering questions was in marked contrast to his demeanor in the Disney
production. He didn't resort to colorful language to appeal to his listeners' sense of wonder or innate desire to explore the unknown; there
were no mentions of a "Columbus of space" or allusions to humanity's
evolutionary leap when first entering the cosmos. From his past experience, von Braun well knew that delivering his message to a room of
grim-faced men wearing campaign ribbons necessitated very different
rhetoric. He framed his talk entirely in terms of adversarial conflict
and achieving strategic advantage and concluded with a note of warning: "The Russians are already hard at work, and if we are to be first,
there is no time to lose."

IN THE SPRING of 1955, the Eisenhower White House approved a proposal for the United States to orbit the world's first satellite during the
upcoming scientific International Geophysical Year. Modeled on International Polar Years of the past, the IGY would involve scientists from
forty-six countries taking part in global geophysical activities and experiments from July 1957 through December 1958. The Soviet Union
was believed to be on the verge of announcing its own satellite program, so the United States was attempting to stake its claim as well.

Very real Cold War concerns lay behind what appeared to be a
purely scientific initiative. Since the beginning of the escalating
nuclear-arms race, the two superpowers had been seeking a method to
monitor the progress of their opponent's weapons programs and their
compliance with international agreements. Rather than using a high-
altitude spy aircraft, which ran the risk of being shot down, the influential global-policy think tank, the RAND Corporation, proposed a
space-age alternative: an unpiloted orbiting observational satellite that
would either transmit video images or return exposed film reels via
automated reentry capsules.

President Eisenhower believed that if a scientific research satellite
was placed in an orbit that passed over the airspace of a Soviet Eastern
Bloc country, this action would establish a precedent for orbital overflight, thus opening up space for observational reconnaissance. But

from a diplomatic perspective this was an unsettled matter of international law. Eisenhower listened to other advisers, who cautioned that if the United States became the first to orbit an object over another nation's sovereign territory, it risked international condemnation. They thought it wiser for the United States to hold off being first.

Von Braun's secret Project Orbiter proposal was vying against two alternative plans, developed by the Air Force and the Navy. Orbiter was generally acknowledged to be the most sophisticated, practical, and reliable of the three proposals, but it was the Naval Research Laboratory's Project Vanguard that got the official nod in late July 1955 when the White House made its announcement. (The projected launch wouldn't occur until late 1957 or 1958.) Von Braun believed Project Orbiter had lost out to the Navy specifically because he had been involved. He wondered whether professional jealousy, his recent celebrity status, and prejudice about his German past, combined with the ever-present interservice rivalry, had contributed to his satellite's rejection. Von Braun was so sensitive about reports of antagonism against him in Washington that when he learned that Disney's publicists were planning to promote a rebroadcast of "Man in Space" with an ad line suggesting that the TV show led directly to the White House's recent action, he vehemently implored them not to do so, fearing that some in Washington would assume he was using his Hollywood connections to take credit. "The statement would hurt the cause far more than it would help," he told Disney's producer.

In fact, a more important factor was involved in the selection. Giving the project to the Naval Research Laboratory was intended to deemphasize military and strategic implications, since the lab already had a reputation for conducting basic scientific research, unlike the Ordnance Guided Missile Center at Redstone Arsenal. The Navy planned to use modified research rockets with an established pedigree from past scientific-research experiments. In contrast, the Army intended to use a modified Redstone, which had been expressly developed to carry a nuclear explosive.

A memo written by Eisenhower's secretary of defense, Charles Wilson, clarifying the nation's missile-development responsibilities, only made matters worse for von Braun's team. Wilson gave the Air Force responsibility for all future land-launched space missiles and limited the Army's work to surface-to-surface battlefield weapons with a range of less than two hundred miles. At Redstone Arsenal and in the city of Huntsville, morale plummeted.

Four days after the White House's announcement of Project Vanguard, Soviet space scientist Leonid I. Sedov addressed reporters at his country's embassy in Copenhagen, where the annual congress of the International Astronautical Federation was being held. Much as the White House had expected, Sedov announced that the Russians would possibly launch a satellite in the next two years. When asked by a journalist about any German rocket specialists employed in Russia, Sedov adamantly insisted that none were working there.

Sedov's assertion was largely correct, despite published rumors to the contrary. The full story didn't emerge for decades. Joseph Stalin had personally authorized that nearly five hundred of the remaining rocket engineers from Peenemünde be forcibly transported to Russia in 1946. They were confined to isolated communities with little more than standard amenities, where they were subject to constant surveillance and the enmity of their Russian counterparts.

The Soviet Union's closest equivalent to von Braun was Sergei Korolev, a Ukrainian-born engineer and strategic planner who, despite having spent six years in one of Stalin's gulags, had risen to prominence as the chief rocket designer. Korolev distrusted the German engineers living in Russia, fearing that if they attained positions of importance they could possibly thwart his own ambitions and undermine the authority of his handpicked team of designers. Nevertheless, by the mid-1950s, and after wasting years working on projects of little importance, most of the captured Germans were finally allowed to return to the West.

A RINGING PHONE in the darkness immediately woke him. Arthur Clarke reached for his glasses and rose from the bed in his Barcelona hotel room to catch the call. The man's voice on the other end of the line said he was calling from London. It was a journalist from the Fleet Street tabloid the *Daily Express,* requesting a quotation in reaction to the breaking news. The reporter informed Clarke that minutes earlier the Soviet news agency, *Tass,* announced that Russia had successfully launched an artificial moon weighing 184 pounds into orbit around the Earth.

It was approaching midnight on October 4, 1957. Clarke was in Spain to attend that year's International Astronautical Federation conference, and the call from London placed him in a privileged position among those who had arrived. Generalissimo Francisco Franco's regime maintained an international news blackout; nothing appeared in newspapers or on the radio until it had been cleared for public release. Most of the other conference delegates remained ignorant of the Soviet Union's triumph until those arriving late brought the news of the outside world with them. But within a few days everyone knew a new Russian word, "Sputnik," meaning "fellow traveler." The provocative response that Clarke gave to the reporter that night received attention in newspapers around the world as well: "As of Saturday the United States became a second-rate power."

Clarke's words did not sit well with William Davis, an American Air Force colonel. "I heard that and I didn't like it! Space is the next major area of competition. If this one is lost, we might as well quit." Colonel Davis suggested the United States should counteract the Soviet propaganda by readying its own space vehicle, with a human on board.

Missing from the Barcelona conference was Wernher von Braun. It was still Friday evening in Huntsville when a British journalist called his office at the Army Ballistic Missile Agency, the successor to the Ordnance Guided Missile Center. Von Braun had feared this day would come, and his sense of disappointment was mixed with anger. He had no doubt that had the United States government given him the opportunity, he would have placed the first artificial satellite in orbit.

At a press conference a few days later, President Eisenhower congratulated the Soviets but dismissed the idea of a race with Russia. He emphasized that both superpowers were engaged in an international program of scientific research. Eisenhower attempted to explain the Soviets' success by casually noting that "the Russians captured all the

German scientists at Peenemünde," a remark that astounded many in Huntsville, as it was an outright lie. Unstated by Eisenhower was his relief that Russia had established the controversial precedent of orbital overflight. In the future, neither the Soviets nor other nations could voice their diplomatic objections when the United States eventually orbited its planned surveillance satellites.

The American media's reaction to Sputnik was far less measured than the president's. Columnists bemoaned the shocking loss of national prestige and criticized Eisenhower's blasé reaction as lack of leadership. Opposition politicians appeared on television, expressing fear that the Soviets might use their powerful rockets to position a nuclear sword of Damocles above any nation.

Overshadowed by the news of Sputnik was another event at the Barcelona astronautical conference that captured the Cold War zeitgeist. Dr. S. Fred Singer, a young Austrian-born physicist noted for his work in cosmic radiation, was there to deliver a provocative paper. One of von Braun's Project Orbiter colleagues and a leading member of the American Astronautical Society, Singer had also cultivated a talent for getting his name into print by fearlessly saying things his more prudent scientific colleagues would avoid.

Singer's paper was titled "Interplanetary Ballistic Missiles: A New Astrophysical Research Tool" and argued in favor of exploding thermonuclear bombs on the Moon as a way to conduct scientific research. It was an idea that, he said, was "not only peaceful, but also useful, and, therefore, worthwhile." This proposal, he believed, might lead to other experiments, such as exploding thermonuclear devices on the planets or an attempt to create a new star. But more immediately, Singer suggested, this initiative would promote world peace, as "the H-bomb race between the big powers would then be reduced to the much more tractable problem of seeing who could make the bigger crater on the Moon." He was especially excited by the possibility of rearranging lunar geography. "The idea of creating a permanent crater as a mark of man's work is an appealing one," he said. "One is left with a nice crater on the Moon which is unnamed and therefore provides unique opportunities for perpetuating the names of presidents, prime ministers, and party secretaries."

The Soviet delegates sitting in the lecture hall listened to Singer's proposal with understandable astonishment and outrage. However, Singer later remarked that the Soviet reaction to his paper was "blown out of proportion."

As Sputnik dominated the headlines, newspapers gave less attention to the big story out of Little Rock, Arkansas, where a week earlier President Eisenhower had ordered 1,200 members of the Army's 101st Airborne Division to assist with the desegregation of Little Rock Central High School. The day before Eisenhower's order to mobilize the troops, more than a thousand white protesters had rioted to prevent nine black students from attending the school. After Sputnik was launched, Radio Moscow seized upon an opportunity to shame the United States for hypocritically calling itself "the land of the free": It alerted its global listeners to the exact moment when the satellite would pass over Little Rock, news that was specifically intended to be heard in the emerging independent nations of Africa.

Incensed that he and his new boss at Huntsville's Army Ballistic Missile Agency, General John Medaris, hadn't been given an opportunity to ready a modified Redstone rocket to launch a swift response to Sputnik, von Braun covertly made his case in the media, despite having received orders from Washington not to make any public comments. Magazine features called him "The Prophet of the Space Age" and "The Seer of Space," portraying him as the brilliant visionary whose bold ideas had been ignored by petty bureaucrats, unimaginative military officials, and cowardly politicians. And knowing that nothing motivated people as powerfully as fear, von Braun evoked the specter of atomic annihilation, arguing that it was imperative that the United States establish its superiority in space if the nation was to survive.

Eisenhower became increasingly irritated by von Braun's arrogance, celebrity status, and self-serving pronouncements. In fact, the president's growing frustration with von Braun likely accounted for his wildly inaccurate comment at his press conference about the Russians having all the German scientists. If it had been intended to belittle von Braun's reputation and public profile, it backfired badly.

The media panic only increased when in early November the Soviets orbited Sputnik 2. A much heavier satellite, it carried the first living creature on a one-way trip into orbit, the photogenic husky-terrier mix, Laika. NBC's Merrill Mueller, a reporter who had covered D-Day, the Battle of the Bulge, and the bombing of Hiroshima, grimly faced the television camera as he told viewers, "The rocket that launched Sputnik 2 is capable of carrying a ton-and-a-half hydrogen-bomb warhead."

Democrats with eyes on the 1960 elections, and possibly the White House, added their dire voices to the chorus of doom. Senator Henry "Scoop" Jackson announced that this was the last chance to save Western civilization from annihilation, while Senate majority leader Lyndon Johnson compared Sputnik to a cosmic Pearl Harbor, warning the nation that we "must go on a full, wartime mobilization schedule."

Taking von Braun's lead, Johnson forecast a sinister future vision that would delight the heart of a James Bond supervillain: "Control of space means control of the world," he declared. "From space, the masters of infinity would have the power to control the Earth's weather, to cause drought and flood, to change the tides and raise the levels of the sea, to divert the Gulf Stream and change temperate climates to frigid."

Eisenhower scoffed at all the apocalyptic rhetoric, as well as at those who offered a variation on Fred Singer's idea that the United States should respond to the Soviet Union's presence in space by sending a rocket to the Moon armed with a warhead. In response, Eisenhower said he would rather have one nuclear-armed short-range Redstone rocket than an expensive and impractical moon rocket. "We have no enemies on the Moon," he declared.

On Capitol Hill, Lyndon Johnson invited von Braun and General Medaris to testify before a Senate preparedness subcommittee inquiry. Johnson not only recieved the media exposure he desired, but the pair from Huntsville put the White House on the defensive with a few well-crafted lines for the newsreel and television cameras. "Unless we develop an engine with a million-pound thrust by 1961," Medaris warned, "we will not be in space—we will be out of the race!" Von Braun raised the specter of a hammer and sickle hanging in the heavens, cautioning that the country was "in mortal danger" if the Soviets conquered space. "They consider the control of space around the Earth very much like, shall we say, the great maritime powers considered the control of the seas in the sixteenth through eighteenth centuries. And they say, 'If we want to control this planet, we have to control the space around it.'"

The country heard them testify to how the Army's readiness had fallen victim to petty armed-service rivalries, bureaucratic lassitude, and indecisiveness, a situation that Johnson called "nothing short of disgraceful." Von Braun once again hypnotized the press, particularly a *New York Times* journalist who described him as a "blonde, broad shouldered and square-jawed . . . youthful-looking German scientist" who "drew many sympathetic laughs as he smilingly grappled with questions."

The nation's attempt to rid itself of Sputnik anxiety came on a morning in December 1957, barely two months after the shocking news from Moscow. Journalists at Cape Canaveral all had their binoculars focused on Launch Complex 18. There had been no official announcement, but word had spread among the newsmen that this was the likely day; sources at the local motels, restaurants, and bars all said something was being planned for that morning.

Shortly before noon on December 6, a cloud of white smoke appeared at the base of the U.S. Navy's Vanguard rocket as it began to move upward into the sky. CBS News's Harry Reasoner observed the launch from a privileged position on the porch of a nearby beach house. At the first sign of smoke he shouted, "There she goes!" His assistant, who was inside the house, on the phone with the network's New York newsroom, immediately conveyed the word. Once the message had been received, Reasoner's New York colleague promptly put down the phone to get the news on the air. But by that moment Reasoner was shouting, "Hold it! Hold it!" as he watched the Vanguard fall back on the pad and collapse into an expanding fireball, its tiny satellite toppling out of its nosecone. CBS had beaten ABC and NBC in broadcasting the news from the Cape but had incorrectly reported that the launch had gone well.

The shock of the failure was somewhat reduced by the fact that pictures of the explosion hadn't been broadcast on live television. But the humiliating film of Vanguard's end was shown repeatedly later in the day, often in slow motion. In the days that followed, the Air Force

Missile Test Center chose to impose tighter media regulations, going so far as to prohibit binoculars and cameras on the nearby beaches.

In the wake of the failed launch, General Medaris and von Braun struggled to find their opportunity. Their Jupiter rocket, built in four stages so as to carry a satellite into orbit, was assembled later that month. But when it came to scheduling an available launch date at Cape Canaveral, the Huntsville team had to compete with their rivals, preparing a second Vanguard. The Navy rocket went through four different countdowns in January, but all were canceled due to technical difficulties or bad weather. The Army was alloted only three days —January 29, 30, and 31—to get Jupiter off the ground.

High winds canceled any possibility of a launch on the first two days, but at the last available opportunity, at 10.48 P.M. on January 31, 1958, Jupiter lifted off. NBC News used a motorcycle courier, a light airplane, and a police escort to rush their film of the launch to an affiliate station in Jacksonville, where it was processed and broadcast nationally within ninety minutes.

By the time the film was ready, von Braun, University of Iowa astrophysicist James Van Allen, and the Jet Propulsion Laboratory's William Pickering were holding a midnight press conference in a small camera-packed room at the National Academy of Sciences in Washington. Pickering's JPL, in Pasadena, had designed and built Jupiter's payload, the Explorer 1 satellite; Van Allen was in charge of the United States's International Geophysical Year satellite program. Basking in triumph before a bevy of reporters, the trio held above their heads a full-size replica of Explorer, in a pose that became an iconic press photo.

In Huntsville, few were watching television at that late hour. The town that had seen its population triple in the eight years since the Army brought its rocket-and-missile center to northern Alabama had informally renamed itself "the Rocket City." Led by a call from its mayor broadcast over the local radio station, Huntsville's police sounded squad-car sirens and honked horns as citizens gathered downtown to celebrate with cheers of vindication and jubilation. Close to midnight, thousands gathered in the town's Courthouse Square, which had been watched over by a granite statue of a Confederate soldier

The Jet Propulsion Laboratory's William Pickering, physicist James Van Allen of the University of Iowa, and Wernher von Braun celebrate the launch of Explorer I, the first American satellite, at their midnight press conference in Washington, D.C., on January 31, 1958.

since 1905, erected and dedicated a few months after a notorious lynching on the same spot. The crowd held signs reading SHOOT FOR MARS!, OUR MISSILES NEVER MISS!, and MOVE OVER MUTTNIK!

The mayor joined the rowdy celebration, lighting skyrockets and ducking exploding firecrackers. A *Life* magazine photographer pointed his Speed Graphic camera at the gleeful crowd, which hoisted high an effigy of former defense secretary Charles Wilson; images with unsettling reminders of the town's past. After his 1956 memo restricting the Army's missile program, Wilson had become the most hated man in Huntsville, and he was now widely blamed for allowing the Soviets to be first in space. As the effigy was torched, the revelers waved American and Confederate flags and shouted, "The South did rise again!"

Von Braun was the man of the moment. His face was on the covers of *Time* and *Der Spiegel*. Eisenhower invited him to a White House state dinner. When he appeared in Washington, overflow crowds of journalists and photographers would attend, yet never were there any questions about his war years. Congressmen requested that he pose for photographs with members of their family. As a Capitol Hill session on space appropriations was breaking up, one congressman was even heard asking, "Dr. von Braun, do you need any more money?"

Von Braun signed with a speakers' bureau and commanded as much as two thousand five hundred dollars for an appearance. Columbia Pictures and a West German studio commenced discussions to determine whether his personal journey from Nazi weapons engineer to Cold War American hero might serve as the basis of a successful dramatic film.

The attention and adulation given von Braun and his Huntsville team neglected one important German without whom Explorer would never have orbited. Hermann Oberth had joined the German rocket community in Huntsville, after von Braun had reached out to his former mentor and offered him work in the Army Ballistic Missile Agency's research-projects office. But Oberth hadn't been able to stay current with the ever-changing technology, and as a German citizen, he was restricted from reading classified information. He worked alone on projects of his own design, but his research produced little of consequence. Oberth's prickly personality didn't help. He felt out of place in Huntsville and in the United States and harbored resentment for some of his former colleagues, who he believed had betrayed him.

The Army also had reasons to avoid bringing attention to the mentor of the world's most famous rocket designer. Oberth was notorious for making provocative or controversial remarks. He might, for instance, insist with icy certitude that at age sixty-four he should be chosen as an astronaut. "They should send old men as explorers. We're expendable." Or he might try to argue why Hitler "wasn't all that bad" or explain how if Germany had won the war Hitler would have funded space travel more vigorously than either the Soviets or the Americans.

And then there were the UFOs. By the mid-1950s Oberth had announced that the rash of UFO sightings indicated that Earth was being visited by extraterrestrials. He appeared at UFO conferences and expounded with deadly seriousness about the lost continent of Atlantis and its connection to Germany.

So when he reached the mandatory retirement age of sixty-five, few in Huntsville objected to his decision to collect his pension back in Germany. Oberth announced he would devote his time to philosophy. "Our rocketry is good enough, our philosophy is not."

VON BRAUN'S TRIUMPH with Explorer did little to rectify the ongoing competition among the different branches of the armed forces. Former defense secretary Wilson's decision two years earlier to give the Air Force responsibility for long-range ballistic missiles implied it would become the branch designated to oversee any future military activities in earth orbit. However, von Braun and General Medaris at the Army Ballistic Missile Agency continued to work on their own ambitious ideas, including plans for a large heavy-lifting rocket that could place human-piloted vehicles in orbit.

The ongoing competition extended into the services' public marketing campaigns, with the Army, Air Force, and Navy each promoting their leadership in the dawning space age. The Navy regarded space as a new ocean to conquer and command, in keeping with von Braun's allusion to the great maritime powers of the past. The Air Force argued that space was an extension of the conquest of the air, just at a higher altitude, and coined its own marketing term, "aerospace." And the Army, while regarding rocketry as a high-powered extension of the artillery, also used the launch of Explorer to promote itself as the team that got things done.

Shortly after Sputnik and the panic on Capitol Hill, the Air Force inaugurated its own piloted spacecraft program, called Man in Space Soonest. Its Special Weapons Center even commissioned a top-secret fast-track study, code named Project A119, to evaluate the scientific paybacks of Fred Singer's proposal to explode thermonuclear weapons on the Moon, an idea some in the Air Force believed would demonstrate to the world America's military prowess and instill patriotic pride at home.

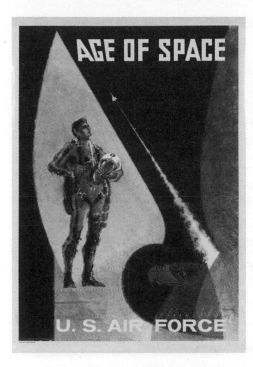

By the late 1950s, the Army, Navy, and Air Force were each employing space age–themed marketing campaigns to encourage new recruits. The Army's poster celebrates the launch of Explorer I, the Navy's includes a picture of Vanguard, and the Air Force promotes its plans for a human military presence in outer space.

President Eisenhower realized he needed to resolve the ongoing service rivalry that was becoming counterproductive and costly to the country. He was also increasingly wary of the power of some personalities in the American military to influence public opinion and gain congressional backing for their ambitious and expensive projects. Accordingly, Eisenhower decided to reduce the military's role in future human spaceflight by signing into law the National Aeronautics and Space Act. His announcement in late July 1958 followed a Presidential Science Advisory Committee that recommended developing space technology in response to the "compelling urge of man to explore and to discover, the thrust of curiosity that leads men to try *to go where no one has gone before.*" It was a declaration that—when slightly reworked with the adverb "boldly" in the prelude to the television series *Star Trek* eight years later—would become one of the most familiar catchphrases of the latter half of the twentieth century. Presciently sensing the emotions those words would invoke, Eisenhower noted in an accompanying letter, "This is not science fiction . . . every person has the opportunity to share through understanding in the adventures which lie ahead."

The National Aeronautics and Space Act created a new *civilian* space agency, the National Aeronautics and Space Administration (NASA), built in part from the half-century-old National Advisory Committee for Aeronautics (NACA), which was already dedicating half its resources to space-related projects, including Vanguard and the X-15 suborbital space plane. Chosen as NASA's first administrator was the president of Case Institute of Technology, T. Keith Glennan, a former member of the Atomic Energy Commission.

Within the first week of NASA's creation, the Air Force terminated its nascent Man in Space Soonest initiative. NASA would now oversee a new civilian-run program, named Project Mercury, dedicated to putting the first Americans into space. Instead, the Air Force would concentrate on its own piloted winged space glider, known as Dyna-Soar, which, after being launched on top of a ballistic missile, would allow military crews to service satellites, conduct aerial reconnaissance, and possibly intercept enemy satellites.

Despite the idealistic rhetoric about exploration and adventure, it was impossible to conceal the reality that the civilian agency planned to send Americans into space atop repurposed military missiles developed to deliver warheads and transport reconnaissance spy satellites. The United States therefore chose to emphasize the open, peaceful, and cooperative nature of its civilian space program, which stood in contrast with the secretive and militarily aligned Soviet effort.

Remarkably, the Soviet Union had never placed a high priority on launching the world's first artificial satellite. Rather, its military rocket program had been developed to inform the world that Russia had the capability to strike other nations with nuclear weapons. Sputnik was an unexpected dividend after von Braun's Soviet counterpart, Sergei Korolev, developed a heavy-lifting rocket—the R-7—capable of delivering a six-ton nuclear warhead. But the Soviet warhead turned out to be far lighter than the original estimate; Korolev had designed a rocket much more powerful than needed.

Korolev realized the R-7 could put a satellite in orbit, if that was of interest to the Kremlin. Following Eisenhower's International Geophysical Year announcement, Korolev sent a memo noting that should Russia want to set a world record by launching a satellite, they could do so at practically no additional cost. When Khrushchev gave his consent, he never anticipated the alarmist reaction in the United States. While the Soviet space program's principal purpose was—and remained—military, Sputnik's overnight success suddenly elevated the role of space research in the eyes of the Kremlin, making it an engineering, scientific, and propaganda priority.

NASA's charter specifically restricted it from any responsibility for military defensive weapons or reconnaissance satellites. That separation between NASA and the Pentagon allowed it to act as Washington's public face for promoting scientific research, furthering exploration, and bolstering national prestige, while deflecting attention from ongoing military space initiatives. In fact, during NASA's first year of operation, the Pentagon's space budget was nearly 25 percent larger.

NASA's formation left the fate of the Army Ballistic Missile Agency in Huntsville uncertain, with both General Medaris and von Braun taking a predictably negative view of the new civilian agency. In addition to the existing NACA facilities, NASA brought the Army's Jet Propulsion Laboratory and the Navy's Vanguard group under its umbrella. Glennan, NASA's new administrator, proposed bringing half of von Braun's group at the Army Ballistic Missile Agency within NASA, an idea that von Braun immediately killed by using his celebrity status to sway political and public opinion. Both General Medaris and von Braun feared that under NASA, the Huntsville group would lose its unique position and become a small part of a larger, and probably dysfunctional, government agency.

As 1958 came to a close, President Eisenhower surprised the world with his own space-propaganda stunt, intended to deliver multiple messages. One was a message of peace; a second was a bit of blatant saber rattling directed at the Soviets; a third was a ploy to silence Eisenhower's critics; and a fourth was a sly dismissal of America's premier space-age celebrity. Under tight secrecy, an Air Force Atlas missile weighing more than four tons was launched into orbit. Transmitted from a tiny box inside the huge Atlas, a recording of Eisenhower's voice proclaimed, "America's wish for peace on Earth and goodwill toward men everywhere." By orbiting the entire Atlas missile—an achievement of little distinction in itself—the United States could technically claim to have placed the heaviest satellite into space.

Still, it wasn't lost on the Soviets that the sentiments voiced on the tape had in fact been delivered by a new ICBM, specifically designed to transport a thermonuclear warhead. Nor was it lost on von Braun and many in the media that Eisenhower had excluded the Army's team in Huntsville from a starring role in a space-age first, intended to boost American prestige. The head of the new Advanced Research Projects Agency, which oversaw Project SCORE as it was called, described it as essentially "a propaganda ploy designed to put a really big, heavy object into space as a means of silencing press and congressional complaints about small payloads and rocket failures."

The media gave the stunt plenty of coverage during the holiday season, but it ultimately achieved little of importance, not even as the first device to transmit a message from space to people on Earth. It was quickly forgotten. Von Braun had concluded that despite their potential to revolutionarily change society, communications satellites would never engage the public or motivate politicians to fund a massive space effort like the piloted program he envisioned. He knew that the fear of annihilation, the loss of military superiority, or the erosion of national prestige were greater motivators among those in power, just as they had been in his dealings with the Third Reich. He was intrinsically aware that the public's imagination would be fully engaged only when an intelligent being was on board a spacecraft, providing a vicarious adventure of a singular and historic nature.

And, indeed, nothing NASA undertook during its first year captured the public's attention more than the selection of the nation's first astronauts. It was this step that brought the dream that had consumed the minds of Clarke, Ley, von Braun, and many others for nearly three decades to the verge of reality.

Sensing that this was not only a good story but also a turning point in human history, journalists searched for a way to portray the men who would experience the unique, dangerous, and otherworldly. Even though there was little to distinguish the first seven astronauts from any other group of military pilots when NASA introduced them at a Washington, D.C., press conference in April 1959, the media rapidly promoted them as exemplars of American masculinity, courage, resourcefulness, and intelligence.

Though they spoke of wanting to travel into space in the near future, both Clarke and von Braun were already a few years older than John Glenn, who at age thirty-seven was the oldest of the seven Mercury astronauts. He had been chosen from an initial group of five hundred applicants, with the finalists representing the Navy, Air Force, and Marine Corps. Five had experienced air combat in World War II or Korea, and the same number had been military test pilots. All were reported to have IQs greater than 130, and, coincidently—or not—all were firstborn or only children. Coming from a fraternity of combat fliers and jet jockeys, the Mercury Seven, as they came to be called, had no intention of being treated like confined lab rats in a glorified orbiting science experiment. They saw themselves first and foremost as active pilots.

However, neither they nor NASA were prepared for how their fairly routine press conference would become a pivotal moment in the marketing of the American space program and the transformation of modern celebrity. Preceded by stories of past air heroes like Charles Lindbergh and a decade of Hollywood science-fiction films, the seven pilots were thrust into starring roles in the television age's first heroic real-life narrative. They and their families were abruptly placed under the modern media's spotlight. Ghostwritten and sanitized versions of their lives appeared in heavily promoted issues of *Life* magazine, the result of a controversial NASA-approved contract that gave the magazine exclusive rights to the personal stories of the astronauts and their wives, even though the men were government employees. Since information related to their work had been understood to be freely available to all, journalists from rival publications naturally felt as though they were being shut out and they criticized the arrangement, to no avail. As part of the agreement, the astronauts were given a life-insurance policy and additional income to supplement their modest military salaries. Almost as important to them, the exclusive nature of the contract gave them justification to decline countless other media requests, and as a result it indirectly protected the privacy of the astronauts and their families when they were not the focus of a *Life* feature.

Not surprising given the tenor of the time, NASA gave no consideration to women as astronaut candidates. Despite the example of Jacqueline Cochran, who had set a series of historical firsts during a career as a military and air-race pilot, no woman in the United States had been granted an opportunity to gain experience as either a combat or test pilot. But provocative speculation never hurt sales, so the question of whether a woman might fly in space, and when, often arose in popular magazines. In reality, it was a non-issue, and NASA avoided public statements that would only exacerbate controversy.

When *Real,* a publication that advertised itself as "the Exciting Magazine for Men," had considered the question in 1958, it concluded that women would indeed have a place on lengthy future space missions—as crew members willing to ease the strong sexual urge of men in the prime of life. The author, Martin Caidin, who became a fixture among the Cape Canaveral press corps during the next decade, attempted to make his argument by considering the alternative: "If you ignore the problem, you're letting yourself in for emotional dynamite and homosexuality—and that is not acceptable." In a similar vein, whenever von Braun was asked a question about the possibility of women serving as astronauts, he usually responded with a wry smile and a prepared answer: "We have talked about adding provisions in the space capsule for one hundred twenty pounds of recreational equipment."

The seven smiling men pictured wearing sports shirts and crew cuts in the pages of *Life* magazine quickly eclipsed the celebrity of America's most famous rocket man. When, shortly after their press conference, the astronauts visited Huntsville to see the rockets under development, von Braun said publicly that he found them wonderful people, "serious, sober, dedicated, and balanced." But behind the scenes during the visit, von Braun and General Medaris were still trying to determine the fate of the Army Ballistic Missile Agency.

Von Braun's newest project was the Saturn, a cleverly improvised design for a large, heavy-lifting booster. Built from existing component parts, the huge rocket's first stage was composed of a cluster of eight individual cylindrical Redstone rocket-size fuel tanks—each eighty feet high and five feet in diameter—surrounding a single, slightly larger Jupiter rocket tank. Five of the tanks, including the Jupiter tank in the center, carried liquid oxygen; the remaining four carried kerosene. Together, the Saturn's six engines would produce 1.5 million pounds of thrust, enough to place a payload of ten thousand to forty thousand pounds into low earth orbit. While the Department of Defense and the Advanced Research Projects Agency had been planning large reconnaissance satellites, von Braun was thinking of other possible uses. He knew that if he could obtain funding to produce a small yet very powerful heavy-lifting booster and demonstrate its ability, the decision makers in Washington were more likely to approve the design of the next, slightly larger model. By progressing in steps to bigger and more powerful vehicles, he would eventually produce one capable of taking men to the Moon, an option von Braun was always working toward, even though no one in Washington was talking seriously about such an undertaking.

NASA's civilian man-in-space program was planning to use the military's Redstone, Atlas, and Titan missiles for the early missions, but any ambitious later projects involving a space station or leaving earth orbit would require a bigger heavy-lifting rocket. Von Braun's Saturn now seemed the likely workhorse for NASA's longer-term future. With Eisenhower's consent, the Army's entire rocket development-operations division in Huntsville was brought under NASA's umbrella as its rocket-development facility. Renamed the George C. Marshall Space Flight Center, it became the agency's largest facility when the transfer took place on July 1, 1960. For the first time in more than a quarter of a century, von Braun would no longer be working for a branch of the military. And his lifelong ambition to design the rockets that would take humans into the heavens was now a reality.

While von Braun's future with the Army was still under discussion, his past was being recreated on a Munich movie-studio soundstage. Columbia Pictures was producing *I Aim at the Stars,* a dramatic biopic intended to tell von Braun's odyssey from a rocket visionary in Nazi Germany to an American hero. Playing von Braun was German actor Curd Jürgens, a familiar face from other recent Hollywood productions. British filmmaker J. Lee Thompson had intended the movie to address questions about the social responsibility of a modern scientist and what constitutes a war criminal, but these moral issues were lost in a screenplay that focused on telling an inaccurate and sanitized version of von Braun's life. When the film opened in London, Munich, and New York, protesters handed out "Ban the Bomb" leaflets and displayed placards denouncing von Braun as a Nazi. However, the most bruising attack came from movie critics, and the film disappeared from movie theaters just as the final days of the 1960 presidential election were playing out on home screens. By the decade's end, 88 percent of American homes had televisions. In addition to the novelty of the nation's first televised presidential debates, the 1960 election marked a turning point in American politics, as the power of the image proved as crucial as the candidates' spoken words.

This was the first presidential election in which both candidates had been born during the twentieth century. Both had also served with distinction during World War II—one returning as a war hero who had saved lives. For those who had fought in the North Atlantic, in Europe, in the Pacific, and in Korea, the election of 1960 marked a dramatic generational shift.

Watching the campaign from Huntsville, one World War II veteran saw something of himself in John Kennedy. Like the Massachusetts senator, von Braun had been born to privilege and wealth and, with a combination of charisma, intelligence, and persuasive rhetoric, had risen to national prominence. Hungry for a change after the cautious policies of the Eisenhower White House, von Braun thought Kennedy might be the right person to usher in the dawning age of human space travel. Kennedy was not afraid of making bold decisions, such as his choice to ignore the advice of campaign strategists and help secure the release of Martin Luther King, Jr., from an Atlanta jail cell during the final week of the campaign.

Von Braun and his wife went to their local Huntsville polling place on Election Day and cast their ballots for the Democratic presidential candidate.

As his second term was nearing its end, President Eisenhower was determined to deliver a final message to the American people. He had been contemplating the content of his farewell address for nearly two years and had labored over more than twenty drafts before appearing in front of live television cameras three days before the inauguration of the thirty-fifth occupant of the executive office. Considered by many the most important speech of his presidency, Eisenhower's televised farewell famously voiced a warning about the increasing influence of the American military-industrial complex. In it he expressed respect for scientific discovery and the ways in which technology could improve lives, but he called equal attention to the "danger that public policy could itself become the captive of a scientific-technological elite."

Eisenhower subsequently declined to elaborate in public about what specifically led him to make this speech. But some months later in a private conversation, a noted nuclear physicist asked the former president whether he had anyone in mind when he mentioned the scientific-technological elite.

Eisenhower answered without any hesitation. He had two people in mind: physicist Edward Teller, the "father of the hydrogen bomb," and Wernher von Braun.

CHAPTER THREE

THE NEW FRONTIER
(1961–1963)

John Kennedy conveyed a sense of confidence and ease as he strode to the podium. The youngest man ever elected president, Kennedy had not yet been in office for three months. Seated before him in the large State Department auditorium were more than four hundred journalists. Present in the room as well were three very large TV cameras. Today's presidential news conference would be broadcast on live network television, as had become the custom in this new administration. Never attempted in the Eisenhower years, these afternoon exchanges between the president and the press were a new attraction, occurring nearly every other week, preempting afternoon soap operas and game shows.

Kennedy's apparent comfort before the cameras that afternoon gave little indication that the days ahead would define his presidency and greatly affect the course of the twentieth century. But April 12, 1961, had begun with extraordinary news. That morning the Soviet Union had announced the successful launch and return of the Vostok 1 spacecraft, carrying cosmonaut Yuri Gagarin, the first human to enter outer space and orbit the Earth.

The news was electrifying yet not entirely unexpected. For the past few days American intelligence sources had been predicting that the Russians might attempt such a feat. A few hours before the news conference began, viewers in Europe had seen the first live television images ever broadcast from inside the Soviet Union. A somewhat blurry transmission from Moscow presented a carefully orchestrated display in which Gagarin stepped from an airplane and strode across a red carpet toward a jubilant and beaming Soviet premier Nikita Khrushchev. The two men were then seen proceeding in a motorcade to a massive celebration in Red Square.

In Washington, however, journalists were even more concerned with another subject: Cuba. Following the revolution two years earlier that had overthrown dictator Fulgencio Batista, American relations with the government of Fidel Castro had deteriorated badly. The United States broke off all diplomatic ties with Cuba only days before Kennedy was inaugurated. In one of his first actions, the new president approved a CIA plan for an armed invasion of this island nation, which he was assured would be carried out in such a manner that American involvement would remain entirely hidden. But in early April, newspapers reported that the United States was training an anti-Castro military unit in a remote Guatemalan base. This prompted the first question from one of the assembled journalists, asking Kennedy not about the Soviet launch but rather how far the United States was willing to go to

The face of Russian cosmonaut Yuri Gagarin as reproduced on newspaper front pages around the world when he became the first human to travel into space on April 12, 1961. It was likely this photograph that caused Arthur C. Clarke to compare him to a modern-day Charles Lindbergh.

assist an anti-Castro invasion. Kennedy's careful response attempted to define the current situation as a conflict between opposing Cuban factions, with added assurance that under no circumstance would any American armed forces intervene militarily in Cuba.

Considering the tension of the moment, the second question from the State Department auditorium, about Gagarin's flight, was far easier to answer. Calling it "a most impressive scientific accomplishment," Kennedy announced that he had sent his congratulations to the Soviet Union and reported that the United States had previously expected that Russia would be the first to orbit a human in space. But, he added, NASA had hopes of doing likewise within the year.

However, a few minutes later—after another question about Cuba—a journalist confronted the president, noting that members of Congress and other Americans "were tired of seeing the United States second to Russia in space." He wanted to know if additional spending would result in the United States being able to catch up and surpass the Soviet Union.

"We are behind," Kennedy conceded. "No one is more tired than I am." And then, after alluding to von Braun's Saturn rocket and other very expensive advanced space projects, he said, "We are, I hope, going to go in other areas where we can be first and which will bring perhaps more long-range benefits to mankind." He went on to speak of the importance of developing technology to desalinate salt water to provide cheap fresh water to the developing world, something he admitted was not as spectacular as a man in space or Sputnik but an accomplishment that would dwarf any other scientific feat.

BUT THE LONG-TERM beneficial implications of Russia's achievement were not what caught the world's attention that afternoon. Photographs of the handsome twenty-seven-year-old Yuri Gagarin soon appeared on the front page of newspapers throughout the world. Readers learned that he was the son of a carpenter and a milkmaid from a collective farm. He was described as a modest, five-foot-two-inch, hardworking, fit, quick-witted elite jet pilot, well liked by his comrades. In Ceylon, where he was now living, Arthur C. Clarke saw the photographs of

Gagarin and immediately compared him to a modern-day Charles Lindbergh.

The latest issues of *Time, Newsweek,* and *Life,* with Gagarin on the cover, were appearing on newsstands across the nation as the CIA's collaboration with anti-Castro Cuba forces went into full swing. But the Bay of Pigs operation, as it came to be known, had an outcome far different from what Kennedy had been assured would happen. And when he failed to provide the necessary air support, the invasion turned into an unmitigated disaster. Instead of a rapid removal of the Castro regime, the Kennedy administration faced a diplomatic crisis and an international public-relations nightmare. Even worse, the Soviet Union's leaders had begun to assume America's photogenic new president was callow and weak.

It was in the wake of these two isolated events, separated by less than a week, that the promise of space suddenly arose as a way to dramatically alter the narrative about America's future and its standing in the international arena.

A White House staffer heard Kennedy angrily complaining in the midst of the media frenzy about Gagarin. "If somebody can just tell me how to catch up. Let's find somebody—anybody, I don't care if it's the janitor, if he knows how."

Prior to this moment, human spaceflight had not played a significant role in Kennedy's imaginative or political thinking. Unlike Clarke and von Braun, Kennedy read no science fiction during his childhood. His bookshelf contained the classics encountered by most upper-middle-class boys his age: *The Jungle Book, Kidnapped, Treasure Island, The Arabian Knights,* and *The Story of King Arthur and His Knights.* The American pulp magazines, such as those Clarke bought from the table at the back of Woolworth's, were not what a boy from an elite school like Choate should be seen reading. Rather, during his prep school years, young John Kennedy carried a copy of *The New York Times* under his arm. Reading books about space travel would have marked him as an intellectual lightweight.

Life magazine's White House correspondent observed that the new president had a weak understanding of space policy. His science adviser, MIT professor Jerome Wiesner, recalled that Kennedy hadn't thought much about the issue during his first few weeks in office. While von Braun had voted for JFK, hoping that he might push the United States toward an ambitious spacefaring policy, those who knew Kennedy well had never heard him discuss such thoughts.

By the time of Inauguration Day, 1961, there had already been eleven unpiloted flights to test hardware for the Mercury program—and the success rate was not good. The first flight of the Mercury Redstone the previous November had reached an altitude of four inches before the rocket settled back onto the launchpad. Kennedy's one mention of space during his inaugural address was to suggest that "together we explore the stars," framing such endeavors as a possible area of cooperation rather than competition. At the celebrity-studded inauguration gala, held the night before, John Kennedy and Lyndon Johnson watched from their boxes as the American space program was satirized by comedians Milton Berle and Bill Dana, the latter in character as reluctant astronaut José Jiménez, a skit that became wildly popular a few months later.

Attending the gala was a thirty-four-year-old Chicago lawyer who had arrived in the capital by train a few hours earlier. Newton Minow had worked on Adlai Stevenson's two presidential campaigns and had clerked at the Supreme Court in the early 1950s. Kennedy had asked him to serve as the administration's new chairman of the Federal Communications Commission, an agency in need of reform after a pair of highly publicized scandals.

Like most who joined the new administration, Minow was a war veteran. He had enlisted in the Army at age seventeen and served in the Pacific during World War II. That experience had forced him to grow up quickly under unusual circumstances, a tempering that forged sobering realism with a sense of idealism and a desire to make a better world for coming generations. John Kennedy once spoke of himself as "an idealist without illusions," a description that Minow thought succinctly summarized the generation that came to Washington to work with the new president.

Television was still a young medium, but Kennedy was aware of its power and potential. He told his new FCC commissioner that he wouldn't have been elected without it, echoing the widely held belief that the 1960 televised debates with Vice President Nixon were a decisive factor in his victory. But television was still evolving; broadcast journalism was primarily headlines read from wire-services reports, occasionally supplemented with filmed images. The three commercial networks' national newscasts were only fifteen minutes in length, with reception for ABC available in only about half the country.

On his first day on the job at the FCC, Minow met a senior commissioner on his new staff, an older man named T.A.M. ("Tam") Craven, who was an engineer.

"Do you know what a communications satellite is?" the older man asked.

"No, I don't," Minow replied somewhat sheepishly.

Craven groaned. "I was afraid of that." He had been trying to get Washington interested in them. "It's the one place where we are ahead of the Russians in space," he explained, "but I can't get anybody to do anything about it."

"I'll tell you what," Minow replied. "If you teach me, and if you are right, I promise you I'll get on it."

As someone who'd practically flunked physics in college, Minow was a bit daunted by this new area of study. But he sensed that it might be something important.

Craven suggested Minow read two books by Arthur C. Clarke, *The Exploration of Space* and *The Making of a Moon: The Story of the Earth Satellite Program*. He also told his boss to take a trip to Bell Labs in New Jersey to see the prototype of something they were working on called Telstar, an experimental communications satellite developed with ATT in conjunction with the national post offices of the U.K. and France. The plan was to place it in orbit over the Atlantic Ocean to relay television pictures, telephone calls, and telegraph images from one continent to the other. Minow soon learned that everything Craven had told him was indeed correct and that the Russians weren't working on anything remotely like this. At the height of the Cold War, communications satellites were a technology that had the potential to actually bring countries closer together, providing a dramatic demonstration of how the United States could use space for peaceful purposes.

In the immediate aftermath of Gagarin's flight and the Bay of Pigs fiasco, President Kennedy sent an April 20 memo to Vice President Johnson, directing him to recommend the best long-range American space initiative that would a) decisively surpass the Soviet Union, and b) demonstrably increase the United States's standing in the eyes of the world. Kennedy said he was open to possible options including the construction of a space laboratory, a trip around the Moon, and an actual crewed moon landing. But weighing heavily on the fiscally conservative president was the expense of such an initiative. Estimates of between 20 and 40 billion dollars were being mentioned in the press that week. (Half a century later, this would be the modern equivalent of 140 to 280 billion dollars.)

Even before the news of the Bay of Pigs invasion broke, pressure had been building on Kennedy to counter the Soviet advances in space. NASA associate administrator Robert Seamans told the House's Committee on Science and Astronautics only two days after Gagarin's flight that NASA believed an emergency all-out national effort could get men to the Moon by the end of the decade, maybe as soon as 1967. But Seamans also criticized the Kennedy White House for its recent decision to cut 200 million dollars from the NASA budget for America's piloted space program. News of Gagarin's flight had come as a blow to others at NASA too. Robert Gilruth, the director of the Space Task Group assigned to putting the first American in space, had hoped to schedule the first piloted suborbital Mercury flight for March 1961, which would have beaten Gagarin into space. However, in this instance Gilruth had run into opposition from Wernher von Braun, who wanted one last unpiloted test of the combined Mercury Redstone before an astronaut was aboard. The missed opportunity felt like Sputnik all over again, although this time von Braun had been the voice of caution.

The press also revealed that, despite assurances that von Braun's powerful Saturn rocket had been deemed a national priority, the government was refusing to approve paying any overtime on the project. "If we are going to get to the Moon first, then we are going to have to allow for some moonlighting," an irritated congressman told reporters.

Questions about overtime on the Saturn project arose as Kennedy

returned to face journalists at his next televised press conference, during the aftermath of the Bay of Pigs debacle. In contrast to his performance nine days earlier, the president's delivery was defensive and
uncertain. Even though he hoped to dramatically challenge the Soviet
Union in space, he could make no announcement until Johnson gave
him a solid recommendation and the expense could be justified.

The reporters began hammering him: "Mr. President, you don't
seem to be pushing the space program as energetically now as you suggested during the campaign. In view of the feelings of many people in
this country that we must do everything we can to catch up with the
Russians as soon as possible, do you anticipate applying any sort of
crash program?"

"Mr. President, don't you think we should try to get to the Moon
before the Russians, if we can?"

Newspaper columnists were no less forgiving. The Pulitzer Prize–
winning military editor of *The New York Times* declared in an opinion
piece that there was an essential flaw in the nation's space policy: a
lack of urgency. He bemoaned that Kennedy appeared to be just as
complacent as the Eisenhower administration had been. "So far, apparently, no one has been able to persuade President Kennedy of the
tremendous political, psychological, and prestige importance, entirely
apart from scientific and military results, of an impressive space
achievement," he pointed out, asserting that Soviet space accomplishments had damaged the image of American power abroad. "Only a
decision at the top level that the United States will win this race [to
the Moon]; only Presidential emphasis and direction will chart an
American pathway to the stars."

After receiving Kennedy's memo, Johnson had spent three days in
meetings with business leaders, Capitol Hill politicians, generals, and
NASA officials to assess the feasibility of political, military, and corporate support for such a huge national undertaking. Notable among the
trio of corporate businessmen Johnson had invited to these meetings
was the president of CBS, the network that was to become most closely
associated with television coverage of America's space missions during
the next decade.

Present, but largely silent, was the usually gregarious new administrator of NASA, James E. Webb. Webb's appointment had been one of the last major administration positions announced by the Kennedy administration. Having little formal scientific or engineering training, Webb did not consider himself an ideal candidate for the administrator's job. A veteran New Deal Democrat, he had been absent from Washington during the Eisenhower years, working for the oil industry in Oklahoma. By the time his name was suggested for the NASA post, it was rumored that seventeen candidates had already declined the job.

Webb noted that from the very first hour of the meetings with Johnson, the vice president was pressing to reach a consensus for a dramatic decision that he could give to Kennedy. The space-station idea was quickly eliminated, since there was a reasonable possibility that the final result would never live up to the exalted original concept; budgetary and technical constraints would likely produce something far less impressive than the initial proposal. On the other hand, the moon landing offered no possibility for half measures—either the country accomplished it or it did not. And if it could be achieved, America would demonstrate to the world it could do nearly anything in space.

Landing a man on the Moon would require an ambitious integration of the United States's economy, educational community, and industry. It would be such a massive national undertaking that Johnson's group of business and military advisers believed it was unlikely that the Russians could get there first. Moreover, since getting men to the Moon would require that both the Soviet Union and the United States develop an entirely new generation of heavy boosters, such an undertaking would nullify the Soviets' earlier advantage in missile development. With von Braun's Saturn already under way, this time the Soviets would be forced to play catch up. Von Braun's rocket, coupled with the country's superior industrial base and advanced technology, looked to set the stage for a realistic triumph.

On the third and final day of meetings, Johnson turned to the new NASA administrator and put him on the spot.

"Are you willing to undertake this?" Once again Webb's response was silence. And then Johnson asked him a second time, "Are you ready to undertake it?"

Webb wasn't concerned about developing the engineering and technology to land a man on the Moon. Rather, he was worried that the national uproar surrounding Gagarin's flight would subside and the public's interest would move elsewhere.

"Yes, sir," Webb finally replied. "But there's got to be political support over a long period of time. Like ten years. And you and the president have to recognize that we can't do this kind of thing without that continuing support. The one factor that's come out of all the studies we've made of the large systems development since World War II is that support is the most important element in success. If the people who are doing it really feel they have strong support, you have a much better chance of getting it done."

Leaving the meeting that afternoon, Webb and his associate administrator, Robert Seamans, walked through a parking lot and across Lafayette Square to the Dolley Madison House, the temporary home of NASA's headquarters.

Stopping for a moment in the parking lot, Webb looked at Seamans. "Are you ready to take a contract to land a man on the Moon?"

Seamans thought about it briefly and then nodded. That was it.

Within a few hours Webb got back to Johnson. "I want you to know I'm all for this. I'm ready to go. But I must repeat to you that this will require the long-term support of you and President Kennedy. Otherwise, you will find [Defense Secretary] McNamara and me running like two foxes in front of two packs of hounds—the press and Congress—and we'll surely be pulled down!"

At the same time as these decisive meetings of Lyndon Johnson's group were taking place, the final launch preparations for the first piloted Mercury mission were playing out at Cape Canaveral. Important details about the mission remained shrouded from journalists covering the flight, particularly the identity of the first American to rocket into space. Weeks earlier NASA had announced that the pilot would be either John Glenn, Virgil "Gus" Grissom, or Alan Shepard. Since the introductory astronaut press conference two years earlier, John Glenn had been the media's favorite. However, on the morning of the first launch attempt, the man in the silver pressure suit and helmet was revealed to be Alan Shepard. Glenn was his backup.

The sting of Gagarin's flight, the humiliation of the Bay of Pigs invasion, and memories of the Vanguard disaster three and a half years
earlier had the Kennedy White House worrying about risking further
embarrassment. At home in Ceylon, Arthur Clarke had the same dark
thoughts. "If they kill one of the U.S. astronauts, it will be just like the
Vanguard situation again," he remarked.

Specific details about Yuri Gagarin's flight remained scarce. Even in
Russia, news of the flight had been kept entirely secret until its success was assured. No film or television footage of the launch had been
released, nor did anyone outside of Russia have a clear idea what the
Vostok 1 spacecraft even looked like. Such information was guarded
under a veil of military secrecy for years following the flight. Instead of
pictures of hardware, photographs of Gagarin's smiling face told the
visual story of the Soviet space program to the rest of the world.

For the preceding two years, NASA's public-affairs officer, Paul
Haney, had pushed to open press access to launch information, including allowing for live television broadcasts of the liftoffs. Military policy
prohibited the distribution of any information unless it had been approved, and current protocol embargoed the release of launch information until the event had occurred. This led to embarrassing situations
such as journalists witnessing the explosion of an unpiloted missile
followed by press officers releasing previously approved statements
that flew in the face of what they all had just seen. Haney had been
arguing that this policy made the public-affairs officers look like idiots
and fostered cynicism and derision.

Under NASA's previous administrator, Haney had gotten nowhere
when trying to change the policy. But once Webb arrived, Haney had
been told to take it up with the White House, which he did. The week
of Shepard's flight, Haney picked up a ringing phone in the makeshift
office of his Florida motel room. The voice on the other end of the line
was the president's secretary, Evelyn Lincoln, calling from outside
Kennedy's Oval Office.

White House press secretary Pierre Salinger then got on the line.
"Paul, the president's wondering about the escape rocket on the Mercury capsule." Kennedy wanted to know what were the odds, in the
event of an aborted launch, that Shepard might die as the nation

watched the live broadcast on television. Haney told Salinger the solid-fuel escape rockets had a success rate of close to 98 percent. There was a pause, and a few seconds later Salinger said, "The president says go ahead. Give it a go. See if it works."

Haney threw the telephone up in the air in jubilation and dinged a ceiling tile. The launch would be televised live to the nation. He later credited that moment as the phone call that put the NASA public-information program in business. There was never a signed policy document. When he later tried to obtain one, Salinger told him, "Oh, shit, we don't sign this stuff. You did the right thing."

With that single phone call, the president of the United States had given the green light to the world's first space-age reality-television series. NASA immediately began working with the national television networks to allow live broadcasts of launches from Cape Canaveral, accompanied by audio commentary from the public-affairs office. The networks were unprepared. The mobile units were operating out of cars and campers parked among the coastal flora in the new press area. Camera tripods had been secured on the roofs of the stationary vehicles, and massive cables snaked through the grass.

The morning of Salinger's and Haney's phone call, a tiny one-line item appeared at the bottom of page 30 of *The New York Times*: "MOSCOW, April 30 (UPI)—M. Bobrov, research director of the Astronomical Council of the Soviet Academy of Sciences, hinted today that Russian astronauts had begun training to rocket around the Moon." Lyndon Johnson was compiling his recommendation for the president while someone in the Soviet Union was attempting to give a clear indication that they were already training for a lunar mission. In the panic of that moment in Washington, few wanted to concede that this might be another Soviet bluff, while those pushing for a stepped-up space program saw little reason to voice any skepticism.

Five days later, on May 5, the three American television networks interrupted their morning game shows to broadcast live America's first human space adventure. When CBS, NBC, and ABC broke into programming at 10:22 A.M., Alan Shepard was already lying on his back in *Freedom 7,* situated atop the Redstone rocket where he had been waiting for nearly four hours.

At 10:34 Eastern Daylight Time, the Redstone was seen lifting off the launchpad on live television. On CBS, correspondent Walter Cronkite was providing audio commentary from a station wagon parked in the press area.

In Evelyn Lincoln's White House secretarial office, a portable black-and-white television with extended rabbit-ear antennae had been turned on. As the time of the launch approached, Lincoln interrupted a National Security Council meeting and a group of approximately twelve filed into her office, congregating near the doorway. Closest to Lincoln's desk were President Kennedy and Vice President Lyndon Johnson. To the side stood historian and special assistant to the president Arthur Schlesinger, Jr., assistant special counsel to the president Richard Goodwin, and assistant secretary of defense for international security affairs Paul Nitze, a ring of pipe smoke surrounding his head. Briefly dropping by was the president's brother Robert, who was carrying a sheaf of papers and had a pencil tucked behind his ear.

On the television broadcast, the commentary was provided by "the voice of Mercury Control," Air Force lieutenant colonel John "Shorty" Powers, whose nickname came from his five-foot-six-inch height. Usually seen wearing a radio headset and holding a clipboard, Powers delivered his public-affairs announcements with a sharp staccato inflection that implied that what he had to say was almost as vital to the mission as was the astronaut riding in the space capsule.

As the Redstone lifted off, the three national networks relied on the image from their pool television camera, which shakily attempted to follow the rocket into the sky. In the background could be heard the sound of muted applause. Viewers watching on CBS heard Walter Cronkite shout, "The Redstone got away all right! . . . Go, go!"

At the White House, Jacqueline Kennedy, dressed for a formal event in white gloves and a pillbox hat, was in a room near Evelyn Lincoln's office. Through the open doorway her husband suddenly called to her, "Come in and watch this!" She joined the group and looked on, leaning against one of the secretarial desks.

About one minute into the flight, Powers announced, "*Freedom 7* reports the mission is A-OK, full go," and he made reference to members of the control team also signaling "A-OK." In fact, neither Shepard

Vice President Lyndon Johnson, Special Assistant to the President Arthur Schlesinger, Jr., Admiral Arleigh Burke, President John F. Kennedy, and First Lady Jacqueline Kennedy watch live television coverage of Alan Shepard's Mercury flight in the president's secretary's office in the White House.

nor the control team had used the phrase. It was Powers's own invention, which he employed at least ten times during the fifteen-minute flight. Over the course of the next few years, the expression became ubiquitous in American households and in advertising copy, with many erroneously believing it to be insider astronaut lingo.

The brief flight of *Freedom 7* sent the Mercury capsule on an arced path that brought Shepard 116 miles above the Earth, before heading back to a splashdown 302 miles from Cape Canaveral. Shepard was weightless for approximately five minutes, but in that time he completed a brief test of the spacecraft's systems.

The three television networks interrupted scheduled programming for fourteen different special telecasts that day, including extended thirty-minute recaps in the evening. British television viewers saw a few brief seconds of the liftoff, sent slowly, frame-by-frame, via telephone cable about an hour later. The BBC also had a mobile unit standing by at New York's international airport, where the American television network pool feed of the launch was recorded on videotape and then immediately placed on a commercial jet headed to London.

On America's third-place, also-ran network, ABC, the man on the beat was a new science reporter, thirty-two-year-old Jules Bergman, who during the next quarter century would cover every American crewed space mission for the broadcaster. Along with Cronkite, Bergman became the TV journalist most often associated with the American space program during the 1960s and early 1970s.

Cronkite's patriotic enthusiasm for his nation's first piloted spaceflight was evident. At the moment Shepard took manual control of *Freedom 7*, Cronkite emphasized that this was something Gagarin had not done. And in truth, Cronkite was noting an important distinction between the two programs: As early as 1959, NASA officials had determined that its astronauts would be active pilots of the spacecraft, as it was conceivable that their individual skills would be needed to successfully execute a mission. In contrast, the Soviet designers favored automated systems that prevented the first Russian cosmonauts from taking dynamic control of their spacecraft. Their primary role was as propaganda; there was little for them to do in space other than return home alive.

When reporting on Shepard's flight, Soviet newscasts took pains to emphasize that the United States had only performed a brief suborbital mission, while Gagarin had completed a full orbit of the Earth. But the public response in the United States was anything but subdued. The flight of *Freedom 7* sparked an outpouring of national pride not seen in years. Neither the television networks nor the White House was prepared for the overwhelming reaction. And the White House's new resident was taking notice.

On a sunny morning three days after Shepard's flight, a trio of Marine helicopters landed on the White House lawn. Emerging from the last helicopter were Alan Shepard, his wife, and Lieutenant Colonel Powers, who were greeted by the president and Mrs. Kennedy and Vice President Johnson. As James Webb, the other astronauts, and assorted congressmen looked on in the Rose Garden ceremony, the president awarded Shepard the civilian NASA Distinguished Service Medal.

A few minutes later Kennedy was due to address the annual convention of the National Association of Broadcasters. Newton Minow, who had been actively working to make the first telecommunications satellite a reality, had received a call from the White House, asking him to accompany the president to the event.

While waiting outside the Oval Office, Minow saw Kennedy gesture to him. "I've got Commander Shepard and Mrs. Shepard in my office. What do you think about taking him to the broadcasters' convention?"

"That would be absolutely perfect," Minow told him.

Kennedy then told the Shepards the plan and said to Minow, "You come with me. I want to change my shirt."

In the White House living quarters, Kennedy proceeded to do just that, all the while engaging in conversation with Minow, who was feeling a bit awkward.

"So, what do you think I should say to the broadcasters?" the president asked.

Fumbling a bit and somewhat intimidated, Minow suggested he might want to talk about the difference between the way the United States conducted its missions in the open and the way they were done in the Soviet Union. "In the United States we invite radio and television broadcasters to be there and to provide the American people with an account of what is going on. In the Soviet Union, nobody really knows what happened—whether it was a success or a failure. Everything is hidden. You should thank the broadcasters for carrying the entire story of Shepard's flight."

On the way to the Sheraton Park Hotel, Minow noticed that the president was in an ebullient mood. He was basking in the moment, which had come after weeks of criticism for the Bay of Pigs and Gagarin's flight. The president and the country suddenly felt good about something they had accomplished.

At the hotel there were rousing cheers and applause as Kennedy introduced the nation's broadcasters to someone he described as the country's "number-one television performer . . . [with] the largest rating of any performer on a morning show in recent history."

The station owners and network executives loved it. And then, in an address that Minow believes was almost entirely improvised, Kennedy artfully enlisted the broadcasters as partners in America's space program by using emotion to meld their innate patriotism with their responsibility as the electronic gatekeepers of a free and open society.

"There were many members of our community who felt we should not take that chance," Kennedy told the audience. He was not only answering those who disagreed with his decision to broadcast the launch on live TV but also subtly countering critics of the piloted space program itself, such as former president Eisenhower, academic scientists, and his own science adviser, Jerome Wiesner. "But I see no way out of it," Kennedy continued. "The essence of free communication must be that our failures as well as our successes will be broadcast around the world. And therefore we take double pride in our successes."

Kennedy then asked his audience, whom he praised as the "guardians of the most powerful and effective means of communication ever designed," to fulfill their national responsibility when promoting the country's defense of freedom. As Shepard stood by his side, Kennedy made an emotional appeal to enlist the broadcasters' partnership. Just what he had in mind would be revealed to the nation two weeks later.

Ironically, Kennedy's well-received speech to the National Association of Broadcasters was quickly overshadowed by the address Minow delivered the next day, an open challenge to television executives to deliver on the medium's enormous promise rather than merely providing "a vast wasteland" of entertainment. He urged the broadcasters to expand their offerings by calling upon a wider array of existing American talent, creativity, and imagination and to have the courage to experiment and include diverse voices and points of view. Minow's speech was subsequently called "the Gettysburg Address of Broadcasting."

Along with the famous images of the young president and his family, a number of the most idealistic and visionary speeches ever delivered—like Minow's—define the Kennedy era and continue to stir the emotions of listeners more than half a century later. Kennedy gave only a handful of speeches about space exploration, but in them he captured the essential aspirational vision that served to explain America's space program of the 1960s. And no speech was as pivotal as the one he delivered to an open session of the U.S. Congress on May 25, 1961, which was titled a "Special Message on Urgent National Needs" but was later known as "the moon-shot speech."

Prior to this address, there had been no groundswell of public interest for a dramatic, expensive space venture. In a Gallup poll conducted shortly beforehand, nearly 60 percent of the respondents were opposed to spending billions from the national treasury to put an American on the Moon. Rather, Kennedy was exercising a leadership initiative. He delivered his speech when his approval rating was at an extraordinarily high 77 percent. Kennedy's address was carried on live television at 12:30 P.M., in spite of grumbles from the networks about keeping it under an hour.

Framing the situation of the lagging American space program in terms of competing ideologies—democracy versus communism— President Kennedy requested an additional 7 to 9 billion dollars over the next five years to fulfill the dream that the space advocates had been urging since the early 1930s. "I believe that this nation should commit itself to achieving the goal, before this decade is out, of landing a man on the Moon and returning him safely to the Earth. No single space project in this period will be more impressive to mankind, or more important for the long-range exploration of space; and none will be so difficult or expensive to accomplish." Geopolitics established the agenda and the goal; science and exploration were secondary.

Despite the huge projected expense to the national economy, little vocal opposition followed it. Republican newsletters published the following week were far more critical of the president's other domestic programs. Gagarin and Shepard had ushered in the age of human spaceflight, and in the emotion of the moment the inevitable destiny of the species in the stars went unquestioned.

Although newspapers had already published front-page intimations of the president's moon-program proposal, some at NASA were caught off guard by Kennedy's speech. Robert Gilruth, who had visited the White House after Shepard's flight, first heard news of the speech when he arrived in Tulsa to attend a space conference. He was aghast at what NASA was facing. At their meeting a few days earlier, Gilruth had told Kennedy, "I'm not sure we can do it, but I'm not sure we can't." But Gilruth, only four years older than the president, sensed in Kennedy a youthful recklessness that might have been a factor in his decision. "He was a young man. He didn't have all the wisdom he would have had if he'd been older. [Otherwise] he probably never would have done it."

Cornered by a journalist at the Tulsa conference, Wernher von Braun provided a quotation crafted for American readers: The United States is "back in the solar ball park. We may not be leading the league, but at least we are out of the cellar."

The singular importance of Kennedy's May 25, 1961, speech may appear to place it as an isolated incident. However, it coincided with another major story unfolding at the time. The day before Shepard's flight, the first Greyhound bus carrying thirteen Freedom Riders into the American South left Washington, D.C., destined for Louisiana. Most of them were college-age students motivated by Kennedy's call to service, volunteering to challenge the existing racial-segregation laws.

The Greyhound bus never arrived in Louisiana. Eleven days into its journey, a mob of people—many of them members of the Ku Klux Klan—attacked and burned the bus in Anniston, Alabama, one hundred miles south of Huntsville. The attackers assaulted many of the students, even attempting to burn them alive inside the bus. On May 20, a crowd armed with baseball bats, broken bottles, and lead pipes attacked another Freedom Rider bus in Montgomery, Alabama.

Besides proving to be a decisive moment in the emerging American civil rights movement, the attacks on the Freedom Riders attracted network journalists equipped with the new lightweight 16mm news cameras that had become available only in the past year. The great news stories of the 1960s would be covered differently than before. Portable 16mm film cameras were taken into space, to the jungles of Southeast Asia, and to the streets of Selma, Alabama, to tell the news with an immediacy and emotional impact that previous generations had never seen.

Had it not been for television, many contend, the civil rights revolution would not have occurred when it did. "When television viewers saw dogs being unleashed against kids who were marching for freedom and saw the violence on their screens, the conscience of America was awakened," explained Newton Minow. "And that had a lot to do with changing public opinion." The images from the South were also affecting America's reputation as a beacon of freedom for the rest of the world. This proved particularly problematic in the newly independent, non-aligned countries of Africa, Latin America, and Asia, which after decades of colonialism were becoming of strategic and economic interest to the two opposing Cold War superpowers.

An addition to the Kennedy administration roster was a government outsider recognized by most American television viewers: journalist Edward R. Murrow, who for more than two decades served as the voice and conscience of CBS News. By 1960, Murrow had grown increasingly disaffected with his relationship with the network and its chairman, William Paley. The CBS chairman's chilly reaction to a Murrow speech criticizing television's failure to deliver on its potential, and Murrow's rivalry with the younger Walter Cronkite, prompted the veteran newsman to accept Kennedy's invitation to oversee the United States Information Agency (USIA), charged with shaping America's public image abroad.

During mid-1961, as the images from Cape Canaveral and Alabama filled newspapers and newsreels, Murrow wondered if negative stories of racial prejudice might be countered by a more inspiring space-related story that would appeal to foreign readers. Murrow typed out a quick memo to his fellow Carolinian James Webb, which asked: "Why don't we put the first non-white man in space? If your boys were to enroll and train a qualified Negro and then fly him in whatever vehicle is available, we could retell our whole space effort to the whole non-white world, which is most of it."

And then he sent a second copy to Robert Kennedy. At the bottom Murrow added, "Hope you think well of the attached idea. It is practically an orphan."

Webb had also been thinking about how powerfully space-age imagery might affect public opinion and support. While he agreed that positive benefits might result from Murrow's proposal, he thought it difficult to implement. It could also enrage some of the staunch segregationist congressmen representing the Southern states that had NASA centers. Nevertheless, Webb added that he would keep it in mind, likely hoping this would be the last he would hear of the matter. It was not.

ALMOST IMMEDIATELY AFTER Kennedy made his landmark address to Congress, he began having second thoughts. The expense was enormous. And after flying a single astronaut on a fifteen-minute flight, no one at NASA had a clear idea how they would get a crew all the way to the Moon and back.

In early June, less than a month after his announcement, Kennedy flew to Vienna for his first—and what proved to be only—meeting with Soviet premier Nikita Khrushchev. There was no formal agenda, but the discussions did not go well. By their conclusion Khrushchev left the meeting convinced that his Western adversary was as naïve and reckless as his recent actions had suggested. But largely overshadowed by subsequent events was a brief discussion about space that had taken place between the two leaders. Kennedy had made a bold and unexpected overture to Khrushchev, proposing that in a public gesture of cooperation the two superpowers combine their resources and venture to the Moon jointly.

At first it appeared that Khrushchev was open to the idea, but the next day he rejected it. When he returned from the summit, Khrushchev explained to his son Sergei that if the two superpowers worked together, keeping Soviet secrets would be impossible. The United States would discover that the Russian ICBMs were far less efficient than they had claimed. Kennedy's proposal went nowhere, but he refused to abandon the idea.

The president's request for more than a half billion dollars for accelerated space exploration was approved by Congress and signed into law the day of the second Mercury Redstone mission, piloted by Gus Grissom, on July 21, 1961. Within days of its passage, advertisements began to appear in trade publications. The defense and aerospace contractor General Dynamics enticed engineers and scientists "just a cut above the average" to consider relocating to San Diego. The ad featured an illustrated montage showing a smiling middle-class couple engaged in water-skiing, tennis, swimming, and golf, promising the im-

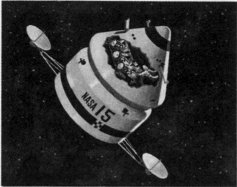
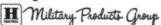
After President Kennedy's space budget was approved by Congress in mid-1961, American aerospace contractors ran trade advertisements to lure additional engineers and scientists. Honeywell touts the coming Apollo program with an early visual conception of the three-man moon ship.

measurable rewards of working on the country's space program while raising a family in Southern California's resort-like climate.

Far more difficult tasks faced the leaders at NASA. Specifically, they needed to determine the best method for successfully accomplishing Kennedy's challenge on time and on budget. NASA had previously undertaken a long-term study of a moon landing as a thought experiment, but the details of how to accomplish Project Apollo—as it had been named in mid-1960—were still very much up for debate.

A decade earlier, in his *Collier's* article, Wernher von Braun had proposed a landing on the Moon using multiple large vehicles assembled near a revolving space station in low earth orbit. A few years later he proposed launching men from the Earth to the Moon in a single gigantic rocket. After jettisoning the large stages that boosted it into space, a single piloted vehicle would then proceed to the lunar surface and subsequently return to Earth, much like the British Interplanetary Society's moon-ship scenario of 1939. In contrast to the "earth-orbital rendezvous" approach utilizing the space station, this straight-line approach became known as "direct ascent." In fact, neither method was practical given the current state of technology in the early 1960s. Additionally, both plans required that hundreds of pounds of fuel and provisions needed only for the final return journey travel to and from the lunar surface with the astronauts.

With Kennedy's 1970 deadline hanging over everyone's head, building von Braun's massive spinning space station for the earth-orbital-rendezvous scenario was too complicated to be a viable option. And while few doubted that von Braun's Saturn rocket could eventually get men to the Moon, the trick of landing on its surface and then launching from lunar gravity to return to Earth added another order of magnitude of complexity. NASA's brain trust next gravitated toward accepting an earth-orbital-rendezvous plan in which a moon vehicle would be assembled in earth orbit from components launched on multiple rockets.

Gradually a third option, which had been dismissed in earlier discussions, was given another look. It was a risky scenario but offered strategic advantages. Known as "lunar-orbit rendezvous," it was championed by John Houbolt, a young aerospace engineer from NASA's Langley Research Center. It proposed launching together on a single rocket two individual small spacecraft—the Apollo and a lunar-landing bug. They would then separate during lunar orbit. The bug would descend to the Moon, land, and then return to lunar orbit. The two craft would rejoin in lunar orbit, the empty bug would be jettisoned, and the entire crew would return to Earth in the Apollo spacecraft. It was an unorthodox proposal as it required multiple rendezvous and dockings far from the Earth, something no one was certain could be routinely accomplished in the weightlessness of space. It also entailed designing separate guidance, environmental, and propulsion systems for both spacecraft, a redundancy many believed would add unnecessary complications.

Von Braun had been a vocal opponent of the lunar-orbit-rendezvous idea when it was first proposed, as it would reduce the role of his group at the Marshall Center. However, he surprised his colleagues at a meeting in mid-1962 when he unexpectedly endorsed the plan. He believed it would balance managerial coordination between Robert Gilruth's group, his group in Huntsville, and those at the Cape, along with the primary contractors, which would eventually include Boeing, North American Aviation, Grumman, and Douglas Aircraft. It was a politically savvy compromise as well, since it deflected some of the rivalry that had been building between the NASA centers.

Simultaneously, another group of engineers from Gilruth's group began considering ways to expand upon Project Mercury, with a second crewed program that would serve as a developmental bridge to Apollo. They coordinated with technicians at McDonnell Aircraft Corporation, the contractor for the Mercury spacecraft, to conceive a larger spacecraft similar to Mercury, dubbed the Mercury Mark II. This soon evolved into a far more complex two-man vehicle renamed Gemini, which would take astronauts on earth-orbital flights for as long as two weeks. These longer missions would pioneer the development of independent electricity-generating fuel cells rather than relying exclusively on batteries and, it was hoped, perfect a method of returning to dry land with the use of a delta-shaped paraglider, forgoing the need of expensive U.S. Navy ocean recoveries.

Few of NASA's debates and long-term decisions about how America would get to the Moon garnered a fraction of the media attention given the immediate human drama of Project Mercury. The genuine popular

curiosity and enthusiasm about Alan Shepard's achievement foreshadowed the next chapter in the escalating international competition between the two superpowers: America's attempt to place a human in
orbit. Realizing this moment would likely generate even greater interest, the network news divisions and their counterparts in the print
media spent the months prior to the first U.S. orbital mission conceiv-
ing how to best tell the story.

Experienced journalists covering America's piloted space program
realized there was an additional aspect to this flight that was markedly
different from the earlier Mercury flights: the astronaut. John Glenn
had neither the laconic wariness of Gus Grissom nor the steely arrogance of Alan Shepard. Glenn's wholesome yet mature smiling freck-
led face was perfect for the television age. He was an articulate,
clean-cut example of American masculinity and seemed the ideal personification of the nation's dominant culture. A genuine war hero, who
had flown sixty-three combat missions in Korea, Colonel Glenn was
already a minor celebrity for pioneering the first supersonic transconti-
nental flight in 1957.

In fact, Glenn had been scheduled to fly a third suborbital mission.
But when the Soviet Union orbited cosmonaut Gherman Titov seventeen times a few days after America's second Mercury suborbital flight,
Glenn's mission plan was changed to an earth orbital trajectory. He
would also be the first human to ride an Atlas, the Air Force's ICBM
missile with a well-known history of launch problems and explosions.

During the early days of the Mercury program, when an actual
emergency occurred it wasn't fully evident to viewers watching at
home. Gus Grissom nearly drowned during the second Mercury flight
while Marine helicopters were attempting to recover his sinking capsule from the Atlantic Ocean, but television viewers were largely unaware of the drama as there was no live television from the recovery
site. A pool journalist on the recovery ship USS *Randolph* transmitted
a live audio report merely relaying the news that Grissom was in the
water but "the concern at the present time is with the capsule." Unknown to the reporter, Grissom had been forced to abandon *Liberty
Bell 7* as it began to sink, and once in the ocean his flight suit had
begun to fill with water. After four minutes struggling to keep his head

above the waves, Grissom was hoisted aboard the helicopter, but his capsule was lost.

Once a rocket disappeared from visual sight after launch, the television networks were forced to resort to their own creative alternatives to visually tell the story of the early space missions. The Mercury orbital flights involved hours of live television coverage, but with no television camera in the spacecraft, producers relied on images of oscilloscopes, console displays, radio antennae, and prerecorded videos of tense flight directors working in the control center. The audio of Shorty Powers's reports from Mercury Control was matched with crude animated images showing the rocket speeding into space and a clock that indicated the elapsed time since launch. It then fell to the network correspondents to convey any sense of actual excitement, since there was nothing to see.

The launch of Glenn's *Friendship 7* was postponed ten separate times before it was finally on its way, and as a result the networks expended hours of valuable airtime in the days leading up to the February 20, 1962, launch. This not only increased suspense but panicked the network accountants. Days before the launch, Glenn's flight was already being referred to as television's "most extensively and expensively reported single news story." Even live gavel-to-gavel broadcasts of the national political conventions required less expense and labor. By early 1962, Walter Cronkite was no longer required to sit in a car while providing his audio commentary. CBS assembled a semipermanent structure, referred to as its "Cape Canaveral Control Center," from which Cronkite could report, seated at a desk with the launchpad visible behind him.

In the White House, John Kennedy and Lyndon Johnson watched the launch of *Friendship 7* on television, surrounded by a few members of Congress. While the room was quiet with the tension of the moment, Johnson turned to Kennedy and said regretfully, "If Glenn were only a Negro."

The country spent the entire day transfixed by the news of Glenn's odyssey as he circled the globe, and by late afternoon millions were following reports of his return. The final minutes of his flight were not without additional suspense: A faulty signal erroneously indicated that the spacecraft's heat shield had detached prematurely. The voice of the otherwise very cool Glenn sounded concerned when he was asked not to jettison the retro-rocket package, an unexpected request added as a safeguard to keep his spacecraft's heat shield secured. Luckily, *Friendship 7*'s fiery reentry was over in minutes and word was soon broadcast that Glenn and his capsule were in the process of being recovered in the Atlantic.

Summing up the nation's psyche on NBC News that evening was Frank McGee, a journalist typically far less prone to showing personal emotion on air than Cronkite. "We have by this time, it seems to me, reached the point of something bordering on hypnosis—irresistibly drawn to the event, and at the same time repelled by our fears of what might happen."

The nation's emotional reaction to Glenn's flight only confirmed what President Kennedy had sensed immediately after Shepard's mission a few months earlier: that television would transform America's attitude toward spaceflight as it had toward politics. And the television networks realized it too, despite the lack of live images from space. When delivering his evening special report recapping that day's landmark flight, NBC's McGee regretfully informed viewers that the network was not yet able to obtain motion-picture film showing the astronaut's recovery from the Atlantic Ocean. At that moment the 16mm film magazine containing that footage was traveling by air to a facility in Florida, where it would be hastily developed and then transmitted to the nation through NBC's New York network feed. Instead, the only image McGee could offer viewers was a grainy black-and-white photo of Glenn on board the recovery ship, which had been transmitted via wirephoto equipment, a technology in use since the 1930s. However, within months, and with little advance warning, television news broadcasting would take a startling leap forward.

FOLLOWING HIS CONTROVERSIAL address to the broadcasters, Newton Minow had become one of the Kennedy administration's most public figures. But more quietly, Minow had been working to circumvent the legal issues surrounding placing the first commercial communications satellite in space and had prevailed upon the president to institute a program for the satellite that would benefit everyone on Earth. NASA would play the role of transportation provider for what would be the first privately sponsored space mission.

Concurrent with Minow's actions, Arthur C. Clarke delivered a paper in Washington on "The Social Consequences of the Communications Satellites." He then traveled to the American Rocket Society's annual conference in New York. During a public panel discussion that included von Braun, Clarke pointed out that if a synchronous television satellite were operational by 1964, it might be possible for everyone around the world to watch the Tokyo Summer Olympic Games. William Pickering, the Jet Propulsion Laboratory's director, was sitting in the audience and was so taken with the idea that he mentioned it to Lyndon Johnson. The next day, when Johnson delivered his keynote address, Clarke's proposal about broadcasting the 1964 Olympics had been inserted into the prepared text.

Audiences didn't have to wait until the 1964 Olympics to see the first live intercontinental satellite television broadcast. It happened only nine months later, while those living in the eastern United States were enjoying a hazy summer afternoon. Viewers who tuned in for the regularly scheduled broadcasts of the soap opera *The Edge of Night* or the game show *To Tell the Truth* discovered that both had been preempted by a special news report. NBC's Chet Huntley, CBS's Walter Cronkite, and ABC's Howard K. Smith alternated hosting duties from studios in New York for a transmission seen not only all over America but in sixteen European countries as well. Two weeks earlier a Thor-Delta rocket had lifted off from Cape Canaveral to place Bell Labs's Telstar 1 in a non-synchronous orbit three thousand miles above the Earth. Cronkite told viewers at the opening of the broadcast, "The plain facts of electronic life are that Washington and the Kremlin are now no farther apart than the speed of light—at least technically," a statement that failed to mention the broadcast was not being seen in Russia. Huntley then introduced a transition to one of President Kennedy's news conferences, already in progress at the State Department's auditorium. With apologies, Huntley cut away two minutes later to introduce John Glenn. Sitting in the Mercury control room at Cape Canaveral, Glenn spoke to the camera about projects Gemini and Apollo before introducing astronaut Wally Schirra, dressed in his silver flight suit.

Unable to view the historic Telstar broadcast was one person who had worked for years to make it happen. During the summer of 1962, Arthur C. Clarke was in Ceylon, though in a nursing home, slowly recovering from a mysterious near-fatal paralysis. His doctors assumed he had suffered a spinal injury as a result of an accidental blow to his head, but decades later, when he was in his seventies, was it determined that Clarke had actually contracted poliomyelitis.

The summer of Telstar, Kennedy was facing a series of difficult decisions regarding NASA's projected 1964 budget. There was a faction within NASA pushing him to increase spending, to eliminate any possibility that the Russians might get to the Moon first, though James Webb was not among them. The president's advisers suggested he conduct a fact-finding tour of NASA facilities and prime contractors to see how all the money was being spent. In St. Louis, the president addressed five thousand workers in a huge assembly room at McDonnell Aircraft Corporation, the prime contractor for both the Mercury and soon-to-be-delivered two-man Gemini spacecraft. He praised their part in "the most important and significant adventure that any man has been able to participate in [in] the history of the world."

A few minutes later, Kennedy looked across the crowd and saw the face of Newton Minow, who had traveled on the vice president's plane.

Kennedy crooked his finger and motioned for his celebrity FCC commissioner to come over, then addressed him privately.

"I understand why Jim Webb and the NASA people are here, but what are *you* doing here?" he asked.

"Well, Mr. President, space exploration isn't only the manned program. I think communications satellites are more important than sending a man into space."

"And just why do you say that?" Kennedy asked.

"Because," Minow continued, "communications satellites will send ideas into space, and ideas last longer than people."

NASA Administrator James Webb looks on as President John F.
Kennedy speaks to journalists during a September 1962 visit to the
space agency's new facility in Houston, Texas. Towering behind
them is a preliminary design of the lunar module, the vehicle the
United States intends to land on the moon before the end of the
decade.

Kennedy didn't respond, but his reflective gaze conveyed to Minow
that he had made his point.

Earlier that day, before arriving in St. Louis, the president had addressed a crowd of forty thousand Texans seated in Rice University's
football stadium in Houston. Speaking in the blazing sun in ninety-
degree heat, he gave a speech that morning that, along with his address
to Congress a year earlier, would become one of the most often-quoted
pieces of space-advocacy rhetoric in history. It was here that he asked
his fellow Americans, and the rest of the world:

But why, some say, the Moon? Why choose this as our goal? And they may well ask, "Why climb the highest mountain? Why, thirty-five years ago, fly the Atlantic? Why does Rice play Texas?"

We choose to go to the Moon. We choose to go to the Moon in this decade and do the other things not because they are easy but because they are hard, because that goal will serve to organize and measure the best of our energies and skills, because that challenge is one that we are willing to accept, one we are unwilling to postpone, and one which we intend to win, and the others too.

Rice University had been chosen as the site of the address because, the previous autumn, Houston had been selected as the home of the new NASA Manned Spacecraft Center, to succeed the facility overseen by Robert Gilruth at the Langley Research Center in Hampton, Virginia. A long list of potential sites in the southern half of the United States had been discussed. However, it was the lobbying effort of Congressman Albert Thomas that managed to make his home district of Houston the winning location for the world's most renowned space-mission control center. Congressman Thomas not only represented Houston and was a Rice University alumnus, but he chaired the powerful House Appropriations Committee, which oversaw NASA's budget. Using his network of connections in Texas business and real estate and at Rice—plus a bit of Washington wheeling and dealing—Thomas arranged that one thousand acres of cattle-grazing land previously donated to Rice by the Humble Oil Company for a federal academic research facility be designated as the location for NASA's new center.

On the day of Kennedy's speech, Rice University was still segregated, in accordance with the wishes of its original founder, who in 1891 established the Rice Institute for the free instruction of white Texans. But if, as planned, Rice University was to partner with NASA's new facility and obtain federal-government funding to create a nearby department of space science, problems lay ahead.

Under the intense Texas sun, Kennedy enticed his listeners with the promise of the future. "Houston, your city of Houston, with its Manned Spacecraft Center, will become the heart of a large scientific and engineering community." What he well knew but didn't mention that day was that his brother, as attorney general, could call upon the Department of Justice to deny any federal money from going to a segregated public institution.

Two weeks after Kennedy spoke at Rice, the university's board of trustees voted unanimously to give the thousand acres to NASA and in a second unanimous vote approved a motion to file a lawsuit to change Rice's charter to admit black students. The trustees' vote was not without controversy; Rice alumni protested, and it wasn't until February 1964 that a jury trial resolved the situation in favor of the trustees.

Kennedy's aspirational rhetoric was not only calling for personal sacrifice to further democracy and freedom and explore the heavens; he was also asking the country to enact social change in order to resolve the effects of the sins of the past. It had only been a year since the pictures of the Freedom Riders' burning bus were seen on television.

As the three-year reign of the seven Mercury space celebrities was nearing an end, NASA held a press conference at the University of Houston to introduce the world to the nine new astronauts slated to fly on the upcoming Gemini and Apollo missions. One of these men might even become the first human to set foot upon the Moon. They were not only younger than their Mercury predecessors, Astronaut Group Two—or "the New Nine," as they came to be known—included the first two civilian astronauts. NASA wanted to make the public aware of this distinction and did so in a savvy and memorable moment of media history. CBS coordinated with NASA to transport a modest middle-aged couple from Ohio to New York, where they appeared as contestants on the prime-time game show *I've Got a Secret.* As the cameras focused on the faces of Viola and Stephen Armstrong, their secret was superimposed onscreen, revealing to television viewers: "Our son became an ASTRONAUT *today.*"

After the secret was guessed correctly following an interrogation by the panel of celebrities, host Garry Moore emphasized that the couple's son, Neil Armstrong, was a civilian pilot, albeit one who had already flown to the edge of space in NASA's experimental X-15 rocket plane. Moore then asked a logical yet prescient question: "Now, how would you feel, Mrs. Armstrong, if it turned out—of course, nobody knows—but if it turns out that your son is the first man to land on the Moon? How would you feel?"

Armstrong was unable to see the program or hear his mother's understated response wishing him "the best of all good luck," his selection having been announced only hours earlier at the press conference. NASA's Manned Spacecraft Center director Robert Gilruth revealed that women had been among the 253 applicants, but "none were qualified." At the long front table with Armstrong were Frank Borman and Ed White, who had attended West Point; James Lovell and Thomas Stafford, who had gone to the Naval Academy; two additional former naval aviators, John Young and Elliot See; James McDivitt, an Air Force pilot who had graduated from the University of Michigan; and Charles "Pete" Conrad, the only astronaut of his generation with an Ivy League education. Conrad, a prep-school dropout with dyslexia (then little known and undiagnosed), went on to graduate from Princeton University with a full Navy ROTC scholarship. Eight of the New Nine would make history in outer space before they were in their early forties.

AFTER A SILENCE of a few months, Edward R. Murrow decided to revive his suggestion that NASA integrate the astronaut corps. He had been repeatedly told that all astronaut candidates must be experienced test pilots, prompting Murrow to write President Kennedy a memo asking whether the United States shouldn't begin training black test pilots now. He added, "The first colored man to enter outer space will, in the eyes of the world, be the *first* to have done so. I see no reason why our efforts in outer space should reflect with such fidelity the discrimination that exists on this minor planet."

At about the same time that Murrow wrote the president, an advisory committee investigating equal opportunity in the military focused its gaze on the nation's most celebrated pilot training school, at California's Edwards Air Force Base. Colonel Chuck Yeager, the man who in 1947 first flew a plane faster than the sound barrier and served as the idol of every jet-fighter pilot in the United States, had just been named commandant of the Air Force's Aerospace Research Pilot School, a

new addition at Edwards. When a White House aide inquired whether
there were any African American pilots enrolled in the ARPS, the answer was a curt "no."

At the request of the White House, the Pentagon began a search to
find a qualified minority candidate for ARPS, ideally a black Air Force
pilot with extensive flight experience and a technical degree. The name
that quickly rose to the top of the relatively short list was a twenty-
eight-year-old Air Force captain named Edward Dwight, who had a
combined total of 2,200 hours of jet-flying time, an outstanding service
record, and a degree in aeronautical engineering.

Opening his office mail on a routine afternoon, Dwight encountered
a letter that was unlike anything he had seen. It proposed he enroll in
the test-pilot school at Edwards as part of a program to be the first
African American astronaut. Well aware of the Kennedy administration's commitment to enforcing desegregation and equal opportunity,
Dwight realized this could be his chance to play an important part in
moving the country forward. But he was mindful that a tremendous
risk accompanied the proposal: If he was successful, he would make
history and a promotion was assured; if he failed, there was likely no
coming back.

Ed Dwight had been fascinated with aircraft since watching P-39
Airacobra fighters fly out of an Army Air Force field in Kansas City during World War II. As early as junior high school, he was borrowing
books from the local library to learn the math and physics required to
pass pilot training exams, study that proved invaluable when he took
an Air Force pilot's exam. After graduating from college he joined the
Air Force, where he learned to fly jets, became a flight instructor, and
earned his degree in aeronautical engineering at Arizona State University.

He was a married father of two children, stationed at Travis Air
Force Base in California, when he followed up on the letter and submitted his Edwards application. Typically after submitting such paperwork months of silence would elapse, but Dwight received an
immediate response, ordering him to Edwards a few days later. He had
expected to be among other African American candidates but discovered he was the only one. Initially, he and the other members of the

new class bonded, unified by their fledging test-pilot status. However, it wasn't long before alpha-male behavior prevailed, with every candidate engaged in a competition of survival of the fittest. Dwight got along well with the other pilots but was also viewed with suspicion. There were rumors that due to his friends in the White House, Dwight had an unfair advantage, and that his success would only come at the expense of his fellow classmates.

Chuck Yeager, Edwards's feisty and notoriously stubborn commandant, did nothing to make Dwight's situation any easier. Yeager had been pressured by Air Force chief of staff General Curtis LeMay to admit Dwight, a request that LeMay told Yeager came directly from Attorney General Robert Kennedy. Believing his school was being used to further the administration's political agenda, Yeager did nothing to hide his resentment. By the early 1960s, Yeager's own fame was being eclipsed by the attention accorded the Mercury astronauts, an elite club Yeager couldn't have joined even if he had wanted to. He lacked both a college education and an engineering degree.

Prior to coming to Edwards, Dwight had idolized Yeager. Once there, however, he encountered a man who exerted strict control over his school and who didn't appreciate outside interference—especially when it originated in the White House. A confidant at the school told Dwight that Yeager had assembled his entire staff of instructors to inform them that the White House had forced him to enroll Ed Dwight in ARPS in an attempt to promote racial equality. Dwight was told Yeager then suggested that if they all failed to speak, drink, or fraternize with him, Dwight would be gone in six months.

Dwight was mindful of the legacy of Major General Benjamin O. Davis, Jr., who endured four years of near-complete social isolation as a member of the West Point class of 1936. Davis, the first African American to graduate from the U.S. Military Academy in the twentieth century, was also legendary as the first commander of the Tuskegee Airmen during World War II. Like Davis, Dwight refused to drop out.

Yeager felt as if his elite school was under siege, especially when lawyers from Robert Kennedy's Department of Justice arrived to investigate claims of racism and discrimination in the Air Force's treatment of Dwight, making an uncomfortable situation even worse. In his auto-

After Captain Edward Dwight was named a candidate at Edwards Air Force Base's elite aerospace pilot school in 1963, many in the press thought it inevitable he would be named America's first black astronaut later that year. However, despite some progress in civil rights made during the 1960s, NASA did not select the first African American astronaut until 1978.

biography written more than two decades later, Yeager likened the experience to being "caught in a buzz saw of controversy . . . [as] the White House, Congress, and civil rights groups came at me with meat cleavers." He explained that if any discrimination was involved, it was based on his conviction that Dwight was not qualified to be in the school.

Despite the adversarial situation at Edwards, Dwight's hard work and determination to persevere paid off, and he graduated sixteenth in his class. But Yeager would only advance his top ten students to the ARPS postgraduate school for astronaut training. When it became apparent that Dwight's name would not be on the list of astronaut trainees, General Curtis LeMay interceded at the behest of the White

House. He made a deal with Yeager to enroll Dwight in the ARPS astronaut school by expanding the number of students from ten to sixteen, a move intended to appease the White House without giving the appearance that Dwight had received any preferential treatment.

Viewers watching a March 1963 NBC Sunday evening newscast heard reporter Robert Goralski announce to the nation that "A twenty-nine-year-old Negro says he is anxious to go into space. He is Edward Dwight of the Air Force, selected to be an astronaut, the first of his race to be so designated. Captain Dwight and his family got the news at his home at Edwards Air Force Base in California."

The press assumed that anyone who finished the ARPS was likely to go into space either as a NASA or Air Force astronaut; however, the manner in which astronauts were chosen remained opaque, rules were often bent and requirements adjusted. Being chosen to attend ARPS did not guarantee future selection on a space mission, but at the time this detail was lost on most journalists not covering the space program.

The national struggle for civil rights was entering a new phase. Alabama's new governor, George Wallace, had been sworn into office with a defiant address proclaiming, "Segregation now, segregation tomorrow, segregation forever!" Dr. Martin Luther King, Jr., and the Reverend Ralph Abernathy of the Southern Christian Leadership Conference (SCLC) were jailed on Good Friday during a campaign in Birmingham to apply pressure on local merchants who practiced segregation. Subsequently, the city's police commissioner approved the arrest of nearly one thousand children during a demonstration and moved against protesters with dogs and fire hoses, producing disturbing images that were printed on newspaper front pages and seen on the national news broadcasts.

In response, entertainers in Hollywood threw their support behind a "Freedom Rally," where Paul Newman and Sammy Davis, Jr., joined Dr. King and a crowd of more than forty thousand at the largest civil rights gathering ever held on the West Coast. Comedian Dick Gregory, who appeared on the stage next to Dr. King, referenced the news about Ed Dwight's selection in his comedy monologue: "I read in the paper not too long ago, they picked the first Negro astronaut. That shows you so much pressure is being put on Washington. These cats just reach back and they trying to pacify us real quickly. A lot of people was happy that they had the first Negro astronaut. Well, I'll be honest with you, not myself. I was kind of hoping we'd get a Negro airline pilot first. They didn't give us a Negro airline pilot. They gave us a Negro astronaut. You realize that we can jump from back of the bus to the Moon?" Gregory also did a riff about landing on Mars and meeting an alien with "twenty-seven heads, fifty-nine jaws, nineteen lips, and forty-seven legs," whose first words to him were "I don't want you marrying my daughter neither."

King laughed along with the crowd at Wrigley Field, but unlike the African American press, which was publishing features about Dwight and his family, King, like Gregory, was no fan of the push for a black astronaut. He saw it as the equivalent of promoting the achievement of an exceptional black athlete on a box of Wheaties. Both did little to further African Americans struggling for advancement. In fact, he thought it could be detrimental, leading those in power to assuage their larger responsibility. "Well, you got an astronaut. What more do you want?"

Shortly before Dwight was to begin the ARPS graduate course, another jet pilot, then an engineer and instructor at Kirtland Air Force Base in Albuquerque, New Mexico, was driving his Volkswagen bus through the New Mexico desert while listening to the car radio. Bill Anders, a twenty-nine-year-old Air Force captain with more than 1,500 hours flying time, had been twice thwarted when trying to gain admission to the Edwards Aerospace Research Pilot School, yet he remained determined to get in.

At the top of the hour, the radio station presented five minutes of news headlines, which included this item: "The National Aeronautics and Space Administration will recruit fifteen new astronaut trainees this summer. The program is opened to both civilian and military volunteers. Cutoff date for applications is July 1, 1963.

"To qualify, a candidate must be a United States citizen born after June 30, 1929, and be six feet or less in height. They must have earned a degree in engineering or physical sciences. Must have acquired one thousand hours of jet-pilot time or have attained experimental flight test status through the Armed Forces, NASA, or the aircraft industry. And be recommended by his present organization."

Anders had all those requirements. But he knew that the mandatory test-pilot experience was coming next, and it would eliminate him immediately.

He listened as the newsreader continued, "Compared to 1962 selection criteria, the maximum age requirement has been reduced to thirty-five, and certification as a test pilot, while still preferred, is no longer mandatory."

Anders was astounded. NASA had dropped the test-pilot requirement. He hadn't given a lot of thought to becoming an astronaut; he just assumed the next logical career move for a hot jet pilot was Edwards's test-pilot school. As he continued listening to the car radio, Anders figured that becoming a NASA astronaut was certainly as interesting an option as becoming a test pilot. He had only three weeks to mail his application.

DURING THOSE THREE weeks, the Soviet Union sent cosmonaut Valery Bykovsky into space in Vostok 5, setting a new endurance record of five days. Attention to Bykovsky was eclipsed two days later when he was joined in orbit by Vostok 6, piloted by a woman. The charismatic Valentina Tereshkova was a twenty-six-year-old former textile-factory worker and amateur skydiver, who was personally selected by Khrushchev to be the first woman in space, a decision conceived entirely as propaganda. Predictably, members of the American press once again asked whether women astronauts might be allowed in the American space program, most notably writer and politician Clare Boothe Luce, wife of conservative Time Life publisher Henry Luce. Surprising many of her anti-communist friends, she criticized NASA's lack of will and apparent sexism while noting that Tereshkova's flight was symbolic of the emancipation of women in Russia, where 31 percent of engineers and 74 percent of doctors and surgeons were female. A year earlier, Congress had held hearings about the astronaut selection process and whether women might be qualified, but nothing had changed. After

Tereshkova's flight, NASA did nothing and again waited for the public discussion to subside.

While Tereshkova got world attention, educators in predominantly African American schools pointed to Ed Dwight's selection as an astronaut candidate as a way to motivate students in mathematics and science courses. Dr. Charles Lang, a science teacher for the Los Angeles Unified School District, sought out Dwight to appear in his educational filmstrip, "Equal Opportunity in Space Science," which was revolutionary for its time. In it Dwight appears with two black schoolchildren, telling them, "Our country is going to need boys and girls with knowledge, imagination, and courage to make it even greater than it is today," and the filmstrip's narrator suggests that the coming space age promises not only "better spaceships" but also opportunities for "creating better cities." Lang gave copies of his filmstrip to NASA's "Spacemobile" educational outreach program, which included it in school science demonstrations across the country. At a time when there were, as Dick Gregory pointed out, practically no black commercial airline pilots employed in America, Dwight's fame prompted requests from public schools, asking him to deliver motivational speeches. "There are no racial barriers for anyone who wants to be a man in space," he told his audiences. "All that counts is whether you can do your job."

Dwight's face appeared on the covers of magazines, not only African American periodicals like *Sepia* ("America Trains First Negro Spaceman") and *Jet* but religious journals such as *The Sign National Catholic Magazine* and *Catholic Digest*. He and his wife were even pictured on the cover of the salacious gossip magazine *Top Secret* ("Integration in Space").

But perhaps the most curious part of Dwight's sudden fame was his part in a moment of international diplomacy, much as Edward R. Murrow had originally envisioned. For the Mercury orbital missions, NASA had established two small African tracking stations, located in Kano, Nigeria, and on the island of Zanzibar. But due to the political volatility of the post-colonial era, NASA had concerns about how long the stations could operate safely. In fact, during the summer of 1963, the State Department warned the technicians in Zanzibar to be on the alert for political rioting and to prepare a personal emergency escape plan. In an attempt to mollify local antagonism about the presence of the American stations, Murrow's U.S. Information Agency had photos of Ed Dwight printed and distributed in Kano and Zanzibar. The hope was that after seeing Dwight's photo, the potential rioters would racially identify with America's first black astronaut and forgo any damage to the American stations.

But the increased media attention only made Dwight's situation at Edwards more difficult. Yeager and Dwight were asked to pose for a series of press photographs in which Yeager assumed the role of the experienced teacher, using a scale model of an F-104 to instruct Dwight about aeronautics. In interviews Dwight gave no hint of the tension he was experiencing. "I'm a pilot. Nobody cares what color a man is here." However, the accounts never mentioned that Dwight was only the third African American test pilot in the history of the Air Force.

NASA received 721 applications for the third group of astronauts. Overseeing the selection board was Mercury astronaut Donald K. Slayton, who had lost his chance to be the second American to orbit the Earth the previous year when he was grounded due to an erratic heart rate. In compensation, he had been named the unofficial "chief astronaut."

In early August, while the selection process was under way, "Deke" Slayton accompanied the second group of astronauts on a five-day desert-survival course in the Nevada wilderness. As part of their training they learned how to fashion desert-suitable clothing from parachute fabric and build a makeshift shelter, which proved exceedingly practical during an hour-long downpour. While in the desert, Slayton was unexpectedly summoned to take an important phone call from Washington. When he returned he told the new Gemini astronauts that the call was from Robert Kennedy, who had requested that NASA accept Ed Dwight as part of the third group of astronauts. "I just spoke for all you guys," Slayton told them as a group under the Nevada sky. "I said if we had to take him and he wasn't qualified, then they'd have to find sixteen other people, because all of us would leave."

Robert Kennedy's call to Slayton came shortly after an extremely contentious meeting between the attorney general, Lyndon Johnson, and James Webb, concerning NASA's poor record of placing African Americans in managerial positions. Believing that Johnson and Webb had talked about equal opportunity but done nothing to effect change, Kennedy decided to call Slayton personally. However, by the time Kennedy placed his call, Slayton and NASA's selection committee had already winnowed down the list of 721 candidates to thirty-four finalists deserving further detailed evaluation.

Despite the competitive situation at the Aerospace Research Pilot School and the way he had been treated there, Dwight remained focused and optimistic. He was confident and believed he would soon be in the elite club. He had been one of only eight men in his ARPS class recommended to NASA "without qualification."

NASA announced that it would hold a news conference on October 18 to reveal the selection of this third group of American astronauts. Advance press accounts indicated that the final selection would number in a range from ten to fifteen and that, of the original 721 applications, two had been from women. However, when Slayton introduced NASA's third group of astronauts, the press saw fourteen white male faces, with an average age of 30.9. Many in that group were to become famous in the years ahead: Edwin E. "Buzz" Aldrin, Alan Bean, Eugene Cernan, Michael Collins, and Dick Gordon. Bill Anders had received the news of his selection in a call from Slayton the day before, on his thirtieth birthday. At the press conference the next day, he remembered thinking, "Why are these people treating us like heroes? All we did was get on an airplane and fly to Houston." He found the experience a little embarrassing. Years later, upon reflecting that many of his colleagues were flying missions over Southeast Asia at the same time he was preparing to go into space, Anders admitted, "Flying on Apollo was much safer than flying over Vietnam."

The questions at the press conference contained few surprises, with reporters asking how the new astronauts' wives felt about their decision ("proud," "supportive," "elated'), their religious preference, and how many had been Boy Scouts. As the press conference was winding down, a journalist directed a question to Slayton about the selection process.

"Was there a Negro boy in the last thirty or so that you brought here for consideration?" a voice from the floor asked.

Slayton leaned into the microphone and curtly stated, "No, there was not."

His answer was followed by a tense silence that filled the room for the next few seconds. Slayton tapped his hand on the table and looked around the room. After a shrug, he glanced at NASA's Houston public-affairs officer Paul Haney and said, "Okay, I guess we're through, Paul."

And with that, the press conference ended. From Dwight's class at Edwards, two were selected: Ted Freeman and David Scott. One would die in a NASA T-38 jet crash within a year, the other would eventually walk on the Moon.

THE KENNEDY ADMINISTRATION'S problems with Cuba and the Soviet Union reached a climax when American intelligence aircraft flying over the Caribbean island discovered photographic evidence of the deployment of medium- and intermediate-range ballistic nuclear missiles. For thirteen days during October 1962, the fate of the world hung in the balance as the United States and the Soviet Union engaged in a nerve-racking battle of wills, which subsided only when the Soviets announced they would dismantle and remove the missiles. The Cuban Missile Crisis brought the world closer to nuclear annihilation than any other moment in the history of the planet and served as a cautionary lesson to both superpowers.

For those who experienced those days in October of 1962, life would never be quite the same. As the crisis developed, the Pentagon distributed a list of likely missile targets in the United States. Upon learning that the Marshall Space Flight Center was on the list, Wernher von Braun, the man who more than any other person spearheaded the technological development of the ICBM, made a decision to construct a reinforced-concrete family bomb shelter behind his custom-built three-level house in Huntsville, Alabama.

Newton Minow hastily departed a New York meeting about communications satellites and rushed back to attend a gathering of the president's Executive Committee of the National Security Council, where he was asked to circumvent Cuba's jamming of the Voice of America radio broadcasts. He devised an emergency strategy with radio stations in southern Florida so that Cubans could listen to President Kennedy when he spoke to the American people about the nuclear standoff. As a ranking member of the administration, Minow was supposed to be evacuated by helicopter to a secure location with the president and the cabinet in the event of a nuclear war. But these plans allowed no accommodations for family members. After sitting through some of the White House strategy meetings, Minow had come to believe the Third World War was likely, and he confided to his wife that he'd already decided not to go if he received an order to evacuate; instead, he would return home to be with his family.

In Minow's estimation, Kennedy's handling of the Cuban Missile Crisis was his greatest legacy. He recalls attending meetings in which the president's military advisers pushed Kennedy to bomb Cuba. Had it not been for the previous mistake of the Bay of Pigs, Minow believes, Kennedy wouldn't have had the courage to overrule his military.

When he chose to address the General Assembly of the United Nations a year after the crisis, President Kennedy wondered if perhaps space might serve as the diplomatic bridge between the two superpowers. After reflecting that sovereignty was not a problem in space, Kennedy asked, "Why, therefore, should man's first flight to the Moon be a matter of national competition? Why should the United States and the Soviet Union, in preparing for such expeditions, become involved in immense duplications of research, construction, and expenditure?" Once again he proposed the idea of a joint U.S.–Soviet cooperative venture to the Moon.

Much had happened since the June 1961 Vienna summit, particularly the missile crisis and the subsequent negotiation and signing of a Limited Nuclear Test Ban Treaty. Kennedy therefore wondered if Khrushchev might view such a joint lunar mission more favorably, in light of recent events. Indeed, there was an American intelligence report suggesting that, contrary to their public statements of the past two years, the Soviets had done nothing to advance development of their piloted lunar program. The presidential campaign of 1964 was looming, and Kennedy's funding for space would likely come under attack. But if there was a way he could leverage the idealism of his cosmic quest to forge a strategic diplomatic partnership *and* reduce the expense of the undertaking at the same time, it seemed a gamble worth placing on the table.

Within NASA offices, Kennedy's surprise UN proposal wasn't met with enthusiasm, but no public dissent was heard. Before addressing the General Assembly, Kennedy had approached James Webb about his suggestion for a joint mission. "I think that's good; I think that's good," NASA's administrator had responded.

Somewhat surprised, Kennedy asked, "That's all right?"

Webb said, "Yes, sir."

Kennedy wanted to make sure that no NASA officials would undercut him by leaking critical comments to the press, and Webb assured him no agency officials would do so. Keeping his word, Webb ordered the NASA centers not to cause any controversy following the UN speech.

With no one from NASA allowed to speak on the record about Kennedy's proposal, *Life* magazine was forced to engage an independent aerospace consultant to explore how such a joint mission might play out. Illustrating the article was a series of commissioned artist's renderings showing a three-person Soviet spacecraft—which looked suspiciously like an Apollo prototype—linking up with an American lunar-landing craft in earth orbit.

Arizona senator Barry Goldwater, who was emerging as one of Kennedy's strongest possible Republican opponents, declared that the United States would need more signs of honesty from the Soviets before signing on to any possible cooperation between the two nations. He added that, if America were to partner with the Russians, those funds would be far better spent on the study of the oceans or increasing agricultural yields than on space.

But from the Soviet Union there was only diplomatic silence. Privately, Nikita Khrushchev discussed Kennedy's proposal with his son Sergei and mentioned that he was leaning toward accepting the idea of a cooperative mission this time. Confused, Sergei couldn't understand why his father had changed his mind. Khrushchev explained that the situation had changed in the last two years, especially after the missile crisis and the test ban treaty. Cooperation in space would give the United States a better understanding of the Soviets' existing nuclear arsenal, which, Khrushchev believed, would naturally lead to reduced tensions and improved relations.

While awaiting Khrushchev's response, Kennedy made a second tour of Cape Canaveral, where he was joined by Webb, von Braun, and two Mercury astronauts. It was during this visit that he witnessed his first and only rocket launch: a Polaris missile breaking through the Atlantic Ocean and into the sky after being fired from the nuclear-powered submarine USS *Andrew Jackson*. Accounts of his mood on that day indicate the visit erased any lingering doubts he may have been harboring about the space program. From the Cape, the president helicoptered to Palm Beach, where he stayed at his father's estate and excitedly described all that he had seen, including a prototype of the new two-man Gemini capsule and the first fully operational Saturn C-1 rocket designed for the Apollo program, scheduled to have an unpiloted test in a few weeks.

The following week, Kennedy spoke under a hot afternoon sun at the dedication of the Aerospace Medical Health Center in San Antonio, Texas. During what was to be his last speech about the space program, Kennedy mentioned what he had seen the previous week.

"Last Saturday at Cape Canaveral I saw our new Saturn C-1 rocket booster, which, with its payload, when it rises in December of this year, will be, for the first time, the largest booster in the world, carrying into space the largest payload that any country in the world has ever sent into space."

The Soviet Union's superior capability to lift heavy payloads into space had been a major concern of the president's ever since his assertions of a "missile gap" in the 1960 election. Looking forward, Kennedy warned against complacency. "We have a long way to go. Many weeks and months and years of long, tedious work lie ahead. There will be setbacks and frustrations and disappointments. . . . But this research here must go on. This space effort must go on. The conquest of space must and will go ahead. That much we know. That much we can say with confidence and conviction."

Recalling a story from a memoir by Irish author Frank O'Connor, Kennedy metaphorically likened the exploration of space to a dauntingly high orchard wall that O'Connor and his childhood friends confronted while venturing into unfamiliar territory. Rather than turn back, the boys took off their hats and threw them over the wall. "They had no choice but to follow them. This nation has tossed its cap over the wall of space, and we have no choice but to follow it," Kennedy said. "Whatever the difficulties, they will be overcome. Whatever the hazards, they must be guarded against.

"We will climb this wall with safety and with speed—and we shall then explore the wonders on the other side."

It was November 21, 1963, and with him at Brooks Air Force Base in San Antonio were Jackie Kennedy, Vice President Lyndon Johnson, and Texas governor John Connally. The next day they would travel to Dallas.

LATE WEDNESDAY AFTERNOON the following week, in his fifth day as president of the United States and his first day in the Oval Office, Lyndon Johnson juggled multiple phone calls. He caught Florida senator Spessard Holland on an airport telephone and then ten minutes later spoke to Florida governor C. Farris Bryant.

Johnson came right to the point with Bryant. "We're getting ready to have the geographical board that controls these park sites rename Cape Canaveral [to] Cape Kennedy."

Bryant's voice was barely a whisper on the other end of the line. "Oh, my word . . ."

Johnson pushed on, giving Bryant a clear indication of why he thought it was important to do this: "So that all the launchings that go around the world will be from Cape Kennedy."

In his call to Senator Holland a few minutes earlier, Johnson had told him the suggestions for renaming the Cape originated in Florida. However, Holland warned Johnson that the Canaveral name went back many hundreds of years and that a study would need to be undertaken before making any changes. Holland said he was sure the local residents would want to feel as if they had a part in making such a change. Johnson assured him he would check this out before taking any action. Johnson's call to Governor Bryant less than ten minutes later clearly indicates he had no such intention.

The next day—Thanksgiving—Johnson addressed the nation from the White House, speaking about President Kennedy and announcing the decision to rename Cape Canaveral, a decision that Holland had correctly predicted was not well received by residents on Florida's northeast coast.

While Johnson told Holland that the renaming idea originated in his state, a White House source the next day told the press that it actually came from Jackie Kennedy, a story she never denied, though she later said the events of that week had been so traumatic she remembered little. She met with Lyndon Johnson in the Oval Office for a half hour the day before Thanksgiving, at which time she mentioned that a fitting tribute to her husband might be something connected to the space program. Before calling the Florida politicians, Johnson also mentioned to journalist Joseph Alsop that he'd discussed renaming the Cape with the former First Lady. In any case, whether Jacqueline Kennedy originated the idea or was talked into it by Johnson during their meeting, she later came to regret it. Johnson, however, realized that branding Kennedy's name on the Florida launch site would indelibly link emotions surrounding the martyred president with support for the space program. For the next seven years, fulfilling the decade-end goal set forth by the thirty-fifth president of the United States seldom went unmentioned when space appropriations and the rationale for going to the Moon came under discussion.

As to Kennedy's proposal for a joint mission with the Soviets, no such reverence toward Kennedy's wishes was heard in Washington. Khrushchev had been preparing a considered response when Kennedy was killed. As a result, the joint mission to the Moon became one of the great "what ifs" of the Cold War era. The day after Kennedy's funeral, UN ambassador Adlai Stevenson alluded optimistically to the possibility of a joint lunar mission in a public speech, and Johnson and Webb briefly discussed the matter during a phone call that week. But the idea went no further. Congress made sure it did not. A month after Kennedy's death, a clause was inserted in an appropriations bill, ensuring that no NASA money "shall be used for expenses of participating in a manned lunar landing to be carried out jointly by the United States and any other country without consent of the Congress."

CHAPTER FOUR

WELCOME TO THE SPACE AGE
(1964–1966)

F. SCOTT FITZGERALD had named the desolate Queens, New York, marshland "the valley of ashes." Beginning in 1910, the tract of land next to the Flushing River had been a dumping ground for ash refuse produced by metropolitan coal-burning furnaces. Within a decade, the Corona Ash Dump, located between Manhattan and the moneyed estates of Long Island, was so large and unsightly that it served as a stark and disturbing reminder of the consequences of unchecked industrial growth and, for Fitzgerald when writing *The Great Gatsby*, a grotesque biblical metaphor.

A quarter century after Fitzgerald's death, the Corona Ash Dump was a fading memory. On that ground in 1964, a ceremony was held to dedicate a very different man-made construction, a towering edifice that rose more than eighty feet above the former dumping ground. Its curvy and cornerless exterior was a single concrete wall broken by a repetitive honeycomb pattern; inside, sunlight filtering through hundreds of stained-glass panels cast the towering interior in an eerie cobalt blue. Visitors described the environment as futuristic or otherworldly. But more often it was described as "cathedral-like."

The Hall of Science at the 1964 New York World's Fair celebrated knowledge of the natural world, the scientific method, and their applications as the nation moved into a highly technological future. It was a secular cathedral dedicated to disciplines many in government, business, and academia believed would define the nation during the decades to come. Unlike the many corporate pavilions, which promoted Coca-Cola, General Motors, Ford, IBM, and other firms, the Hall of Science was one of the few structures to remain after the fair closed in 1965. City officials hoped it would join the new Lincoln Center, the American Museum of Natural History, and the Metropolitan Museum of Art as one of New York City's must-see cultural landmarks.

On the day the Hall of Science was dedicated, present in the minds of many was the belief that science and technology were transforming human beings into a species that would travel to other worlds and find a home in the cosmos. As the nation's foremost government representative of that vision, NASA administrator James E. Webb appeared as the dedication's featured speaker. He was also a vocal proponent for science education as a national imperative. Webb was now overseeing a massive government initiative that was receiving 4.3 percent of the national budget. Most of that money was directed toward fulfilling President Kennedy's geopolitical lunar mandate, but, more quietly, Webb was also attempting to ensure the foundation of the country's scientific, technological, and aerospace manufacturing infrastructure.

In his brief speech, Webb contrasted the emerging new age of applied science with the more mystically informed past: "The world of science is a world of accumulated knowledge, not a world of magic or mystery." Webb asserted that the country's future depended upon the nation's universities and schools ensuring that scientific literacy was an integral part of every education "in the American conviction that public knowledge is public strength." And should his words be misinterpreted as those of a hardcore technocrat, Webb emphasized that science was not separated by some mystique from the humanities but, rather, part of the search for truth, along with philosophy, history, and poetry.

UNLIKE THE PREVIOUS candidates who had been approached for the job, the man President John Kennedy finally engaged to run the nation's space agency in 1961 was neither an engineer, nor a scientist, nor an academic. Thirty years earlier at the University of North Carolina, James Webb had considered a career in science education. However, when at age fifty-four he was named NASA administrator, Webb was an experienced Washington insider, having served as President Truman's director of the Bureau of the Budget and later as his undersecretary of state. Webb's experience in the federal government and in private industry gave him a rare combination of management, legislative, and legal insight. In addition, he was an astute observer of personal interrelationships, a keen listener with the ability to quickly perceive others' concerns and motivate them to do their best. It turned out Webb was the perfect person for the job, even though when he was sworn into office neither he nor President Kennedy had any idea that within a few weeks Webb would be faced with the massive challenge of overseeing a program to put a human on the Moon.

Webb had been raised in a politically progressive home, where love of learning was valued more than the pursuit of wealth. In his early twenties he took a job as secretary to the powerful chair of the House Rules Committee, which provided invaluable insight into the inner workings of Washington. He was fortuitously situated at the fulcrum of legislative action during President Franklin Roosevelt's historic first one hundred days, not only observing how deals were made but actively participating in the process. Two years later his government work with the aircraft industry led to an executive position at Sperry Corporation, an important supplier of military and civilian aircraft-navigation equipment; he eventually served as treasurer, secretary, and a vice president.

In the months immediately before Sputnik, while working as the director of an oil company, Webb created the Frontiers of Science Foundation, an initiative in Oklahoma conceived to train a new generation of secondary school and university science educators and expand statewide technological development. And in the immediate aftermath of the Sputnik panic, it turned out to be remarkably prescient, as the quality of scientific and mathematical instruction in the United States came under increased scrutiny. President Eisenhower gave the Oklahoma program attention in a televised speech in which he emphasized the importance of science education throughout the country.

Now as head of the nation's space program, Webb saw it as a responsibility of his position to motivate and educate the youth of America about the science and technology that would define their future. NASA partnered with the World's Fair Corporation to create a Space Park adjacent to the Hall of Science. There, beneath shadows cast by full-size Atlas, Titan II, and Agena missiles, fairgoers could inspect mock-ups of the new Gemini capsule scheduled to be piloted into space in the coming year. They could also gaze at the Apollo Lunar Excursion Module, which was being designed and manufactured a few miles away, at the Grumman Aircraft Engineering plant in Bethpage, Long Island.

The American effort to get to the Moon was the largest peacetime government initiative in the nation's history. At its peak in the mid-1960s, nearly 2 percent of the American workforce was engaged in the effort to some degree. It employed more than four hundred thousand individuals, most of them working for twenty thousand different private companies and two hundred universities. Webb had to oversee a system that could effectively monitor the various NASA branch offices and contractors, ensure that public and political support for the lunar program was sustained for an entire decade, and keep everything moving ahead, on budget and on schedule, to meet Kennedy's end-of-decade goal with as yet unproven—and sometimes not yet invented—technology. Still, during his tenure at NASA, the man famed in Washington circles for "managing the unmanageable" remained largely invisible to the public. James Webb avoided personal publicity. While overseeing America's quest to the Moon, he was never featured on a single news-magazine cover.

Webb's presence at the opening of the Hall of Science framed NASA's ambitious pursuit of the lunar prize as more than an exercise in global politics. The New York World's Fair's official motto was "Peace through Understanding," but implied in many of the corporate exhibits and displays was a second theme: "Faith in Technology." The optimistic glimpses of the world yet to come reflected the national zeitgeist of the Kennedy years, coinciding with the heady period when humans first entered outer space. It was a time when it was assumed the United States would undertake big challenges and successfully meet them. General Motors's popular Futurama exhibit took visitors on a trip to the year 2024. Highly detailed models depicted modernist underwater cities, massive space stations in rotation as they orbited the Earth, and the first human colonies on the Moon. It was an extension of what Wernher von Braun, *Collier's,* and Walt Disney had forecast a decade earlier.

If rockets to the Moon and the exploration of the planets were to be a part of life in the near future, surely it seemed possible that the twentieth century's other, more earthly challenges could be solved as well. The technological optimism on display at the World's Fair aligned with James Webb's own bigger vision for NASA. He believed that achieving the lunar-landing mandate could serve as an instructive test case to demonstrate how an effectively managed, innovative, large government program could partner with private industry and educational institutions to better America's standard of living and its quality of life. It would lead the way for other large, non-military government programs that might revitalize the cities, establish a mass-transportation infrastructure, reduce environmental pollution, or discover new alternative sources of energy. In this regard Webb was a very different kind of space-age visionary. He wasn't driven by a need to solve a theoretical problem or fulfill a personal dream of traveling into space. Instead, he believed the application of government power and resources could transform the nation through what Webb came to refer to as "space-age management."

By the time the Hall of Science opened, popular forecasts of a greater technological future were coinciding with Lyndon Johnson's outline for his Great Society, in which no child would go unfed and no youngster would go unschooled. When the World's Fair first opened its gates on a rainy April morning, President Johnson marked the occasion with a speech that contrasted the fair's vision of hope for the future with the unavoidable realities of poverty, prejudice, crowded cities, diminishing resources, and the erosion of the country's natural beauty. But at the United States Pavilion, Johnson's words were drowned out by chants of "Freedom Now" and "Jim Crow Must Go" from a small group of well-organized demonstrators. In response, some in the VIP area demanded of the New York police, "Do something! . . . Take them out of here!" before the demonstrators were physically dragged away to patrol wagons. New York policemen arrested more than three hundred activists at the World's Fair demonstrations that morning.

James Webb, Lyndon Johnson, and the protestors who turned out in the rain all believed in the federal government's benevolent power to effect changes to improve society. The dreams, hopes, and expectations for the promising future on display at the New York World's Fair were built on a collective faith in technology. Though difficult political and social issues faced the nation, the mood of the country remained optimistic—despite the recent tragedy in Dallas. Intimations of cynicism and disillusionment about established national institutions were still rare in the mid-1960s. For the moment it seemed possible that innovative space-age management solutions might prove effective when adapted and applied to some of the nation's other daunting challenges.

WHEN HE WAS sworn in to oversee NASA, Webb stood out in an administration populated with younger faces, patrician accents, and Ivy League pedigrees. Solidly built at five foot nine inches, Webb didn't have the athletic, aristocratic presence of Kennedy and von Braun or the towering height of Johnson, but his stance, unflinching gaze, and frequently raised chin telegraphed self-assurance. He spoke rapidly, directly, and often at length, usually inserting a disarming smile and making one-to-one contact with a flash of brilliance in his eyes.

Like Kennedy, Johnson granted Webb near-complete autonomy over the massive budget under his control. However, in contrast to Kennedy, who viewed NASA's focus almost exclusively in terms of beating the Russians to the Moon, Johnson was sympathetic to Webb's belief that within a few years NASA's overall efforts would establish to the world the nation's preeminence in space, not merely the fact that it could accomplish one dramatic feat. In Webb's definition, preeminence included not only the high-profile human missions but also engaging in planetary science with robotic probes, establishing a network of weather and communications satellites, and fostering scientific education and technological innovation.

As a manager, Webb avoided the latest ideas promoted by elite business schools. Rather, he preferred theories and philosophies of administration that he had discovered in practice, often championing ideas that had gone out of fashion or were overlooked. While vice president at Sperry, he explored human-relations-centered management concepts, including the ideas and writings of Mary Parker Follett, a turn-of-the-century theorist, political scientist, and lecturer. Webb referred to Parker as "a brilliant woman" and "high spirit" who emphasized the importance of building group relationships, conflict resolution, and executive leadership that focused on working *with* others rather than exerting arbitrary power *over* them. By the time of her death in 1933, Follett's theories had fallen out of favor, yet Webb continued to mine her work and recommend her ideas while at NASA.

His own leadership at the space agency was marked by an agile "keep-'em-guessing" technique he termed "planned disequilibrium," a process that fused multiple coalitions of forces into a cohesive but essentially unstable whole, which Webb then had to keep headed in the desired direction. Understandably, working under such fluid, stressful, and unpredictable conditions exerted a toll, but many also flourished in this environment, and they tended to be young, competitive, intelligent, and highly motivated.

Overseeing an agency that depended on a workforce largely trained in the sciences and engineering put Webb in a delicate position, given his non-technical background. He resolved this during his formative years as NASA administrator by establishing a decision-making triad with two other space-agency veterans: Robert Seamans, a former MIT engineering professor who was named his associate administrator, and Hugh Dryden, a noted physicist and science administrator who served as deputy administrator. Webb interacted comfortably with engineers, contractors, and the different NASA center directors, but he could occasionally surprise a meeting by asking the participants to consider a more flexible approach to solving a problem. Nevertheless, under Webb's guidance, the individual NASA centers instituted different management systems designed around their needs. For example, at the Marshall Space Flight Center, von Braun communicated with his top managers by circulating a once-a-week compendium, which collected the various managers' concise weekly progress reports supplemented with von Braun's informal thoughts and suggestions.

Out of necessity, NASA and the Pentagon had to work in close collaboration. The launch facilities at the Cape Canaveral Air Force Station were used for the pre-Apollo flights, and in 1963 NASA formally established its own independent Launch Operations Center—later renamed the Kennedy Space Center—adjacent to the military installation. The recovery of NASA's spacecraft also required the active participation of U.S. Navy vessels and aircraft. But it was seldom an easy partnership. In contrast to the Pentagon's embrace of rational systems analysis under the leadership of defense secretary Robert McNamara during the 1960s, Webb's agile and far less rigid planned-disequilibrium approach raised eyebrows. Since NASA's creation in 1958, there had been many in the Pentagon—and in the Air Force, especially—who believed America's space program should be under military control. And after NASA's appropriations grew to exceed 4 percent of the national budget, McNamara's dismissive attitude about the civilian space program led Webb to believe the secretary of defense privately thought, "How [can] anything as big as this be well run unless I'm running it?"

Webb hadn't seen combat during World War II. He was to have been part of the Marine Corps air assault on Japan, which was called off after the atomic bombs ended the war. But he was no stranger to physical risk, having survived four midair engine failures while a young pilot in the Marine Corps reserve. Like most of the war veterans involved in the early space program, he knew that risk was unavoidable whenever a human being was placed on a rocket and hurled into space. Still, he had a keen awareness of the potential repercussions of any public failure. Precisely how much risk Webb deemed acceptable was defined early in his tenure, before the first piloted Mercury flight. Webb clashed with Kennedy's science adviser, Jerome Wiesner, who argued that NASA should spend another year launching primates into orbit before launching the first American astronaut. Webb passionately disagreed. If the Russians were already flying men into space, he said, the United States couldn't be seen wasting time with monkeys. "Our people [have] to learn to live with risk and make damned sure that every effort [is] made to avoid casualties."

Wiesner and Webb passionately disagreed once again, a year later, about the risk involved with the lunar-orbit-rendezvous option. Calling it "the worst mistake in the world," Wiesner believed the plan was a blueprint for disaster. During Kennedy's two-day tour of NASA facilities in 1962, Wiesner and Webb argued about it in front of the press while touring the Marshall Space Flight Center. Eventually Kennedy had to intervene, humorously attempting to defuse the conflict by explaining, "Webb's got all the money and Jerry's only got me."

Overseeing a relatively new government agency accorded a huge portion of the federal budget, Webb took pains to protect NASA's public image from any suggestion of waste, influence, or abuse of power. This extended to his daily behavior. He refused the government limousine assigned to him by the General Services Administration. Instead, he traveled to Capitol Hill in a modest black Checker, similar to the ubiquitous metropolitan taxis of the era. Concerned that Wernher von Braun's lucrative public-speaking schedule might give an appearance that he was profiting personally from his NASA fame, Webb instructed him to drastically reduce his lecture schedule and waive an honorarium whenever he spoke at a university.

Webb's busy schedule took him to NASA's many field centers and contractors' facilities, forcing him to spend much of his working day traveling by air, so, early in his tenure, Robert Seamans suggested NASA replace its aging prop plane with a new business jet. Webb objected. "We're not getting any fancy jets in this organization! As soon as you do that, every congressman who is involved in our program will want to borrow the jet. No jets!" Instead, Webb agreed to upgrade to a slower Grumman Gulfstream I, a widely used twin-turboprop business aircraft, the back of which Webb had outfitted with a small four-seat conference area. This became his airborne office. When heading to one of the field offices or to Cape Kennedy, he might invite an influential congressman along and use the opportunity to subject his guest to a twenty-five-thousand-foot-high persuasion offensive. He would often strengthen his pitch for congressional support by asking a noted NASA scientist or astronaut to accompany them and give an impassioned explanation of why the project in question should be funded.

During the course of his first months at NASA headquarters, Webb realized NASA's public-relations efforts were woefully ineffective, and if public and congressional support were to be sustained for an entire decade they would need to be rethought. Through his North Carolina connections in Washington, he made an appeal to a fellow Carolinian who was already familiar with the space program and NASA's public-affairs efforts.

Julian Scheer had been a respected journalist for *The Charlotte News*, where he had covered the early years of the civil rights movement, the rise of far-right white-supremacist organizations, presidential campaigns, college sports, and the first Mercury launches. But Scheer had recently left his job to write a novel based on what he had observed while covering the struggle for civil rights. He had already co-authored a nonfiction book, *First into Outer Space,* with a young aerospace engineer and was now eager to capture in fiction his impressions of what he considered the other big American story of the 1960s.

Scheer's experience covering the early Mercury flights had been both exhilarating and frustrating. He saw how the television networks and *Life* magazine were receiving favored access from NASA's public-affairs officers, while he and hundreds of other print journalists were

Considered by many the best public-relations person in 1960s Washington, D.C., as NASA's head of public-affairs veteran journalist Julian Scheer worked with the news media to sustain public interest in the country's space program. The author of many books for children, Scheer posed outside NASA's Washington headquarters in 1965, the same year one of his books won a Caldecott Honor Award.

left to cobble together stories working from little more than bland official press releases. And when covering Scott Carpenter's *Aurora 7* flight in May 1962, Scheer had been among the reporters struggling to obtain accurate information from NASA spokesperson Shorty Powers during Carpenter's troubled reentry and recovery. Carpenter had landed 250 miles from his intended splashdown point; however, members of the press were kept uninformed about his status for more than a half hour, uncertain whether Carpenter had survived the fiery reentry. In fact, recovery aircraft had picked up *Aurora 7*'s radio beacon during the descent, and Carpenter's heartbeat monitor indicated he was alive and well. After that frustrating experience, Scheer decided to quit *The Charlotte News* and finally write his novel.

Scheer had been working on his book for only a few days when Webb approached him with a request. Based on everything Scheer had seen and experienced while covering the space program, Webb asked, what kind of a revised public-affairs program would best benefit the public, the press, and NASA? Despite all the fine talk of its open program, Webb believed that what NASA was trying to accomplish for the country wasn't being conveyed effectively. Project Gemini was about to be introduced, and Webb thought this was an opportunity to build upon the public's interest and strengthen support leading forward to Apollo. He envisioned the average American following the nation's Gemini astronauts as they confronted a series of advancing challenges, each of which would have to be mastered before taking the next step to the Moon. Plans called for Project Apollo to follow swiftly after the final Gemini mission. The first flights to the Moon would commence soon thereafter, leading to the eventual lunar landing before the end of the decade. Webb wanted Scheer to think about how his skills and knowledge as a journalist and storyteller might be applied to strengthen public engagement.

He gave Scheer a month to compose a new integrated public-affairs plan. After reading his proposal a few weeks later, Webb immediately called Scheer. "I accept your offer to go to work for me," Webb said. He told him writing that novel could wait. If the job interested him, Scheer would report directly to Webb at NASA headquarters. In return, Webb

promised, he would do everything he could to back up Scheer's strategic plan.

Once installed in his new Washington office, Scheer attempted to
correct the problems he had personally encountered while covering
the Mercury launches. He determined that much of the ineffective-
ness of NASA's messaging led back to Shorty Powers, the ubiquitous
public-affairs officer who had become a minor television celebrity.
When Scheer had Powers removed, Webb did not question the decision, despite the uproar. Upon leaving NASA and the Air Force a year
later, Powers became the first space-age personality to cash in on his
fame, signing a long-term contract with Oldsmobile. Throughout the
mid-1960s, Shorty Powers's second act was appearing in a long-running
series of television ads showing him behind the wheel of an Oldsmobile Jetstar 88, touting the rear trunk's "spacious payload bay" or rev-
ving up the Rocket V8 engine to "take this baby downrange."

With his arrival in Washington, Julian Scheer brought the flavor of a
metropolitan newsroom to NASA's Washington headquarters. To enter
his office was to be surrounded by constantly ringing telephones.
Scheer would juggle calls while reclining behind his messy desk covered in piles of magazines and newspapers. "He'd lean back with a
cigarette curling up into one eye and somehow he managed never to
blink," recalled Brian Duff, who worked with Scheer at NASA. "He'd
say, 'I'll tell you one goddamned thing,' and he'd scratch his stomach at
the same time and you'd think, 'This guy is tough as nails!'"

Smart, witty, irreverent, and media savvy, Scheer was described by
The Washington Post as "the best PR man in Washington." When he
realized that the astronauts were generally uncomfortable confronting
the press, Scheer gained their trust and respect by encouraging them
to be as authentic as possible in public and preventing any attempts to
script their comments.

Among his former colleagues in the press Scheer was well regarded;
however, tensions arose when he believed their work was irresponsi-
ble. Scheer had a particularly contentious relationship with ABC's
Jules Bergman, who would use ruses to obtain network exclusives. In
one such instance, Bergman worked around Scheer's office when at-

tempting to arrange a meeting between Russian cosmonauts and American astronauts at a hotel bar, where his ABC News camera crew was waiting to record the exchange. When Scheer found out, he called the meeting off and made sure that when the astronauts and cosmonauts finally met a few days later at a space conference, the event was open to all the world's press.

Jointly, Webb and Scheer agreed that NASA's open program should send a message to the world: "We're not the Soviets. In America we do things out in the open." Scheer's redefined public-affairs group now made an effort to provide journalists with as much accurate and timely information as could be provided. However, since ensuring continued public support was a central objective, NASA's open program was still far from transparent. Space-beat journalists who saw their role as more than astro-cheerleaders would become frustrated when pursuing investigative stories. Even former journalists working in NASA's public-affairs office occasionally misunderstood the nuance. During one internal presentation, a public-affairs staffer projected a glass slide that read: "NASA has an open program because it respects the public's right to know." Immediately, James Webb interrupted the presentation and demanded, "Destroy that slide." The slide was removed, and shortly afterward the sound of breaking glass was heard in the background as the order was literally carried out. "NASA has an open program because it is good public policy to have an open program," Webb explained. "I'm not in the business of the public's right to know. There are others, like the attorney general, who will take care of that."

Nevertheless, within NASA some resisted Scheer and Webb's more open approach. Engineers and members of the military worried that openness would endanger the secrecy of American technology or open NASA to another possible public humiliation like the failed Vanguard launch. Some of the astronauts, protective of their sanitized heroic public image, feared the press might report they occasionally used inappropriate language and weren't the Boy Scouts depicted in *Life* magazine. Between the agency's efforts to craft a truthful narrative that would sustain the public's support and the press's hunger for drama and exclusives, a tension arose that would be tested repeatedly.

Assured that human space exploration would continue to generate interest well into the next decade, Hollywood producers began looking for viable space-related properties. Inevitably, Julian Scheer's office was approached whenever a film crew requested NASA's technical assistance or permission to shoot on location. Frank Capra, the Hollywood filmmaker renowned for directing *Mr. Smith Goes to Washington* and *It's a Wonderful Life*, had become interested in the American space program while making a promotional short film for an aerospace contractor. He thought it the perfect subject for a Columbia Pictures big-studio feature based on a space thriller written by aerospace journalist Martin Caidin. *Marooned* followed the story of a Mercury astronaut stranded in space when his retro-rocket fails to fire and the emergency-rescue mission that is improvised to bring him home. When Scheer learned the book had been optioned, he tried to prevent NASA from offering any assistance, as the story's plot relied upon the failure of existing technology and the drama was driven by the possibility of an astronaut's death in space. As Scheer expressed in an internal memo, "It would be better for the agency's standpoint if this picture was never made."

A far more promising film project was being planned under the direction of Stanley Kubrick, who had released *Dr. Strangelove* only weeks after Kennedy's assassination. Scheer learned that Kubrick was collaborating with Arthur C. Clarke on an ambitious, optimistic epic about humanity's destiny in space three decades hence. In the preceding year and a half, Clarke had partially recovered from his near-fatal case of polio, though he still suffered lingering weakness in his limbs. But after almost a year of rehabilitation, he was once again able to travel for extended visits to the United States.

Kubrick and Clarke met for the first time over lunch in a Midtown Manhattan restaurant on the same day that the New York World's Fair opened; the conversation lasted close to eight hours. Clarke discovered that while growing up in the Bronx, Kubrick had been an avid reader of *Amazing Stories,* and he shared with Clarke an optimistic belief in science. Clarke was impressed with Kubrick's near-insatiable desire to understand almost everything, from technology to music, and art to philosophy. It was a unique meeting of two avidly curious and creative minds.

A few days after their first meeting, Clarke and Kubrick traveled to Flushing Meadows to see corporate visions of the future. At the General Motors Futurama, they saw models of the rotating space stations and lunar colonies, which they would later reinterpret in the epic film they were already discussing. Clarke had jokingly begun referring to their project as *How the Universe Was Won;* a year later it was given the tentative title *Journey Beyond the Stars;* and finally, when released four years later, it was known as *2001: A Space Odyssey*. It would become not only one of the most celebrated films in the history of cinema but the most fully realized version of the space-age vision that von Braun had introduced in *Collier's*.

As Kubrick and Clarke explored story ideas for their epic during the spring of 1964, they met for long conversations at the Guggenheim Museum and took extended walks through Central Park. Clarke introduced Kubrick to specialists in many areas of space science, including a twenty-nine-year-old astrophysicist from Harvard named Carl Sagan, whose career had been altered when he read Clarke's first book. Clarke also introduced Kubrick to two Marshall Space Flight Center staffers, author Frederick I. Ordway and illustrator and designer Harry Lange, whom Kubrick quickly hired to serve as consultants on his film. At Marshall, Lange had been head of von Braun's future-projects department. His NASA illustrations of imaginary future space vehicles were often used as persuasive visual props whenever von Braun was making a pitch for space funding on Capitol Hill. He once joked that Lange's art was so effective that it actually *produced* money in a town where everybody only spent it. Although he had no experience as an art director for the movies, Lange was responsible for nearly the entire look of Kubrick's epic, from the design of space vehicles and costumes to the spaceship interiors. When framed by Kubrick's lens, Lange's eerie, efficient, antiseptic environment—an extrapolation of the techno-optimism marketed at the New York World's Fair—intimated the film's larger philosophical ideas. As humans ventured into outer space, Kubrick and Clarke suggested, they entered a realm where they would be

Teenage Arthur C. Clarke's interest in the possibility of human space flight was launched by the colorful dust jacket of the British edition of The Conquest of Space, *which he glimpsed in a bookstore window.*

Chesley Bonestell's illustration showing the ignition of the third stage of a ferry rocket forty miles above the Pacific Ocean appeared on the cover of the first space-themed issue of Collier's *magazine in March 1952.*

Wernher von Braun's large Saturn C-1 rocket under construction at the Marshall Space Flight Center in 1961. In the foreground is a Redstone rocket used to launch the suborbital Mercury missions; behind it is the Jupiter rocket.

The seven Mercury astronauts pose in alphabetical order beside a Convair F-106B Delta Dart: Scott Carpenter, Gordon Cooper, John Glenn, Gus Grissom, Wally Schirra, Alan Shepard, and Deke Slayton.

Wernher von Braun's Soviet counterpart was engineer and spacecraft designer Sergei Korolev, shown here on the right wishing cosmonaut Yuri Gagarin a successful flight prior to the launch of Voskok 1 on April 12, 1961.

Two days after he became the first American to orbit the Earth, astronaut John Glenn shows President John F. Kennedy his Friendship 7 capsule outside Hangar S at Cape Canaveral Air Force Station.

President John F. Kennedy attends a briefing during his September 1962 tour of Cape Canaveral. Sitting with him in the front row are NASA Administrator James Webb, Vice President Lyndon Johnson, Director of Launch Operations Kurt Debus, and Secretary of Defense Robert McNamara.

"We choose to go to the Moon in this decade and do the other things, not because they are easy, but because they are hard; because that goal will serve to organize and measure the best of our energies and skills, because that challenge is one that we are willing to accept, one we are unwilling to postpone, and one we intend to win."
President John F. Kennedy at Rice University in 1962.

A human settlement on the Moon in the year 2024, presented as part of General Motors'
Futurama exhibit at the 1964–1965 New York World's Fair.

Astronaut Ed White becomes the first American to walk in space during the June 1965 Gemini 4 mission. He and James McDivitt were the first NASA astronauts to prominently display the American flag on the shoulder of their space suits.

Astronaut Frank Borman prepares for Gemini 7 in 1965. A medical endurance mission, the flight would transport Borman and Jim Lovell on 206 orbits around the Earth during nearly fourteen days. Onboard was a paperback of Mark Twain's Roughing It, but Borman never had enough free time to read more than two chapters.

Gemini 7 photographed from Gemini 6A during the first close rendezvous in space, December 1965. "It was like the Blue Angels at 18,000 miles per hour," said Gemini 6A's commander Wally Schirra. "I did a fly-around inspection. . . . I could move to within inches of it in perfect confidence."

Gemini 8 astronauts Neil Armstrong and David Scott wait to board the USS Leonard F. Mason *after their emergency return from orbit. The original flight plan called for them to splash down in the Atlantic Ocean, but a secondary recovery area in the Pacific Ocean 500 miles east of Okinawa was chosen instead.*

The crew of Apollo 1, Gus Grissom, Ed White, and Roger Chaffee, photographed near Pad 34 in January 1967, days before they were killed when a fire spread through the oxygen-rich atmosphere of their enclosed spacecraft.

The launch of the first Saturn V moon rocket on November 9, 1967. The creation of von Braun's rocket team at the Marshall Space Flight Center in Huntsville, the Saturn V flew thirteen missions, ten with people on board.

With a neighbor providing support, Susan Borman (right) watches the ABC News coverage of the launch of Apollo 8. Her faith in the space program shaken after the Apollo 1 fire, she believed it unlikely her husband would return from the first trip to the Moon.

"Hey, don't take that [photograph]. It's not scheduled." One of the most famous images of the twentieth century was the result of a lucky accident. Known as "Earthrise," this single photograph was worth the cost of the entire Apollo program, said anthropologist Margaret Mead.

Neil Armstrong waves to the press as he leads the crew of Apollo 11 to the van that will transport them on the ten-mile journey to Pad 39A. It was 6:27 on the morning of July 16, 1969. An observer standing a few feet away likened it to "seeing Columbus sail out of port."

From the press site located three miles from Pad 39, ABC News science editor Jules Bergman reports during the live television broadcast of the launch of Apollo 11. The weather that morning was very humid with temperatures already heading into the high 80s.

Prior to the lunar landing, Buzz Aldrin checks out the lunar module while it is still docked to the command-and-service module.

Neil Armstrong photographed in the lunar module after landing on the Moon.

Buzz Aldrin descends the lunar module ladder. NASA scientists contemplated the remote possibility that energy stored in the lunar surface particles might set off a combustible reaction when it came into contact with the astronauts' boots. For that reason Armstrong was told to touch the surface lightly with his heel during his first step.

Aldrin deploys a seismic experiment package during the moonwalk. Less sophisticated than later nuclear-powered Apollo seismometers, this solar-powered device operated for less than a month. It was so sensitive that it picked up the motions of Neil Armstrong turning over in his sleep within the lunar module following the moonwalk.

The iconic image of Aldrin on the lunar surface, with Neil Armstrong reflected in the mirrored visor. Because Aldrin was assigned to take photographs of the landscape and the condition of the lunar module, there are few good images of Armstrong standing on the Moon.

forced to confront their finite biological destiny. Machines with emerging consciousness, like the film's HAL 9000 computer, were far more suited to deep-space missions than were the film's vulnerable human astronauts, whom HAL attempted to eliminate as an exercise of evolutionary superiority. And *Homo sapiens* themselves, when placed in contact with a greater alien intelligence, would transition to a different and superior living entity—not unlike the future of humankind depicted in Clarke's novel *Childhood's End.*

During his two-year-long preparation before the cameras began rolling, Kubrick sent a note to one of his NASA-connected advisers, assuring him that the space agency would be delighted when the film finally appeared on the screen. Kubrick, always averse to revealing details about his films while in production, even invited a member of Scheer's NASA public-relations staff and a small contingent of NASA officials to visit the film set at MGM's studio outside London. Afterward, George Mueller, NASA's head of the Office of Manned Space Flight, jokingly referred to Kubrick's office as "NASA-East" and made a personal request that, once Kubrick had finished, he give him the model of one of the film's spaceships to decorate his Washington office.

Unknown to Clarke, Kubrick, and NASA, on the West Coast another Hollywood writer-producer was outlining a proposal for an ambitious science-fiction adventure series, set on a starship in the twenty-fourth century. It was conceived with the aid of a careful reading of Clarke's most recent book of nonfiction, *Profiles of the Future: An Inquiry into the Limits of the Possible,* which led to thinking about many of the ideas and inventions—like a matter transporter—that were eventually seen in the television series. The same week that Clarke and Kubrick first began exploring their ideas for 2001, Gene Roddenberry sent three copies of his outline and two dollars to the Writers Guild of America to register the name of his series: *Star Trek.*

When it appeared on television in 1966, *Star Trek* was very much a projection of mid-1960s American culture and values, mixing Kennedy-era idealism, Cold War politics, and pervasive sexual stereotypes. It celebrated the superiority of American democratic values while suggesting that other points of view should be entertained. But perhaps the series' most influential message was simply an assurance that humanity would have a future and would eventually prevail over the petty conflicts and divisions of the present day by harnessing the power of technology for the betterment of all.

As Mercury transitioned to the Gemini era and the World's Fair celebrated the new space age, American support for its space program was far less robust than might be assumed from the popular media. Opinion polls at the time revealed that Americans thought favorably about the space program but consistently cited it as one of the first programs that should be cut in the federal budget. Unknown to the American public, CIA intelligence reports indicated that the Soviet Union had no current plans for a crewed lunar-landing program. If these top-secret accounts were correct, the United States was engaged in nothing more than a race with itself. When John Glenn was contemplating a run for the Senate during Lyndon Johnson's first year as president, the former astronaut observed that the threat of possible Soviet domination of space no longer motivated congressmen to unwaveringly approve NASA's budget.

The Soviet space threat briefly resurfaced during the run-up to the fall 1964 presidential campaign, as Lyndon Johnson prepared to face off against Senator Barry Goldwater. When outlining his proposed anti-communist strategy in Southeast Asia, Goldwater criticized the Apollo program as an extravagant waste of money that could be more wisely spent on the military. Johnson, worried that he might be vulnerable to being characterized as "soft on communism," used the occasion of a notable American accomplishment in the space race to return to the panicky rhetoric he'd used during the aftermath of Sputnik.

In the White House Cabinet Room, Johnson welcomed NASA officials to a public event marking the successful conclusion of the Ranger 7 probe's mission. As it zoomed toward the Moon, Ranger transmitted a series of images from its four moonward-facing cameras, creating a dramatic sequence of shots that ended with its final impact. Both Arthur Clarke and Stanley Kubrick stayed up past midnight in New York to watch the live broadcast of the press conference as the Jet Propulsion Laboratory released the first photos. Johnson celebrated Ranger's success by announcing that the United States had now pulled ahead of the Soviet Union in the space race. He referred to Ranger 7 as a "peace weapon," and in a public conversation with a NASA JPL scientist, Johnson warned that if the United States wavered in its commitment to Apollo, the nation's position of leadership in a global battle for survival would be endangered. A leading Democratic House member echoed the clarion of alarm when he predicted that a Goldwater victory would return American space policy to the pre-Sputnik era.

When holding the Oval Office, Lyndon Johnson's rhetoric about the American space program never approached the stirring words of his predecessor, but in many respects his support was even more decisive—dating back to the role he played in Kennedy's decision to go to the Moon. Not only would America's space effort answer the Soviet threat, but Johnson saw it as a benevolent way for the federal government to fundamentally transform the country's economy and foster job opportunities in engineering and manufacturing, particularly in the American South, then so politically important to the Democratic Party. In tandem with these changes, Johnson thought the space program would spearhead government efforts to enforce equal opportunity. In a speech titled "The New World of Space," Johnson announced, "The shackles of Earth are being broken and the resulting freedom will affect us all. . . . Because the space age is here, we are recruiting the best talent regardless of race or religion, and, importantly, senseless patterns of discrimination in employment are being broken up."

In November, Johnson easily won the election, carrying forty-four states and winning 61 percent of the popular vote. During the campaign, Democratic strategists had painted Goldwater as a dangerous extremist, particularly after he incautiously alluded to the possible use of low-yield atomic weapons against North Vietnam. Johnson, on the other hand, had campaigned as someone who could act decisively against foreign aggression, based largely on the actions he took three months before Election Day. In the hours after Johnson had celebrated the success of Ranger 7 in the Cabinet Room, the USS *Maddox*, a United States Navy destroyer on a mission to gather intelligence off the Vietnamese coast, was chased and attacked by three North Vietnamese Navy torpedo boats. This series of events and those that followed—including reports of a second nonexistent attack two days later—were known as the Gulf of Tonkin incident. They ultimately led to the legal justification for American troop involvement in Vietnam and the military escalation that followed. Johnson's actions received widespread public support, and the Gulf of Tonkin Resolution was passed in the Senate with only two opposition votes. However, Johnson's presidential legacy, the future of the American space program, and the ultimate course of American history were irreparably impacted as a result.

THE INTENSE ACTIVITY of NASA's Gemini program—ten crewed flights over less than two years—occurred simultaneously with the escalation of the Vietnam War and some of the most momentous years of the struggle for civil rights. All three stories accelerated the evolution of television news during the 1960s, and how they were visually portrayed on home screens affected the public's emotional perception of the issues.

Gemini was first and foremost a necessary transition to Apollo, which wouldn't be ready to fly until late 1966 at the earliest. The two-man missions would give the astronauts and engineers an opportunity to test and refine equipment and procedures that would be needed later when voyaging to the Moon. These included learning how to navigate, approach, and rendezvous with another vehicle; physically dock with a second craft; experience the effects of long-duration missions of up to two weeks; and venture outside the spacecraft and learn how to work in that environment.

In conjunction with NASA's public outreach, James Webb believed, Gemini would serve as a persuasive strategy to sustain public and congressional support for NASA and the lunar program. With each successive mission, the astronauts would demonstrate a progressive sequence of acquired skills and accomplishments, somewhat akin to an episodic adventure narrative with new installments every three or four months. And nothing was more important when telling the story of Gemini to the world than the photographs brought back from space. The spectacular color images shot during the Gemini missions were

Ten Gemini Titan missions with two astronauts aboard were launched from Cape Kennedy between March 1965 and November 1966. The Gemini program accomplished the necessary milestones before astronauts could be sent to the Moon, including completing successful long-duration flights, rendezvousing and docking with other vehicles in orbit, and mastering an ability to maneuver and work outside the spacecraft.

like nothing that had been seen on any previous flight, and their reproduction in the pages of magazines, newspapers, and on television visually charted the nation's rapid and dramatic extraterrestrial progress.

After the United States's last Mercury flight, in spring 1963, the Soviet Union executed some hastily improvised space missions intended to capture world attention with their novelty, including the first woman in space. Just a few days prior to the first scheduled crewed

Gemini flight in early 1965, the Soviets did it again by broadcasting television pictures of the first person to walk in space. Cosmonauts Alexei Leonov and Pavel Belyayev were launched in a Voskhod spacecraft, a modified Vostok reconfigured to hold two men. Less than two hours into the mission, Leonov entered an airlock and proceeded to perform a twelve-minute spacewalk outside the spacecraft. Russia later released a color motion-picture film of Leonov floating away from a mounted camera fixed to the Voskhod, with central Asia shrouded in clouds one hundred miles below. On the brow of Leonov's space helmet were large red letters reading CCCP, leaving no doubt of the nationality of the world's first spacewalker. The film implied Leonov's feat had been easily accomplished. But the cosmonaut's experience was anything but smooth. The vacuum of space caused his pressure suit to inflate and stiffen, so severely limiting his mobility that he had difficulty maneuvering himself back into the Voskhod. Only by gradually purging oxygen to depressurize his suit did he manage to force his way back through the airlock.

The Soviet Union had no need to accomplish the first spacewalk when it did, other than to snatch propaganda headlines before the United States. A week after Leonov surprised the world, NASA launched Gemini 3. It was America's first piloted flight in nearly two years, but in comparison to Voskhod 2, it was anticlimactic. The Soviets had already orbited three men in the cramped Voskhod 1 the previous fall, so when Gus Grissom and John Young completed three orbits in Gemini 3, there was little to suggest the United States wasn't lagging behind once again, despite Lyndon Johnson's confident words following Ranger 7's lunar mission the previous year.

But less than three months later, everything changed. Gemini 4's four-day mission came close to meeting the Soviet endurance record set in 1963. It also attempted the first rendezvous of two vehicles in space, as command pilot James McDivitt tried to move closer to the second stage of the Titan II missile that had placed it in orbit. (Attempting an actual physical docking would wait until a later Gemini flight.) The exercise soon revealed that maneuvering a piloted craft closer to another object in space was far more difficult than had been imagined and that to do it successfully would require not only radar—

which Gemini 4 did not have—but additional training so the astronauts could accurately calculate the speed and distance differentials between the individual orbits of the two vehicles. Gemini 4 sighted the upper stage of the Titan, but when it tried to move closer, the two objects actually moved farther apart due to their differing orbital velocities.

More significantly, Gemini 4 is remembered for a second event, which also took place on the mission's first day: the dramatic spacewalk executed by pilot Edward White. White was originally scheduled to merely open his hatch and stand up in the cockpit as a test of the external suit and life-support system. But a review of the Soviet film of Leonov's feat encouraged planners at Houston's Manned Spacecraft Center to add a full spacewalk to Gemini 4's flight plan, less than a month before its June launch date.

The person assigned to carry out the American space program's first extravehicular activity—or EVA, as this and all future excursions outside a spacecraft became known—was ideally chosen for the task. White was the "astronaut's astronaut," an archetype who inspired not competitive envy but near reverence; he was an exemplar who reminded others of their aspirations. The son of an Air Force major general, White was renowned among his peers for his physical strength and athleticism; at West Point he almost qualified for the Olympic track team. Novelist James Salter knew White when both served as Air Force pilots in West Germany during the mid-1950s. He described White as "a man who could be relied upon—in every way" and someone whose handsome presence convinced those around him that they were "intimate with greatness." Salter had no doubt that White would leave his mark on history.

Ed White stepped out of the Gemini spacecraft while it was speeding at 17,500 miles per hour over the Pacific Ocean, as the West Coast of North America came into view. Anchored to the Gemini by a twenty-five-foot gold-plated tether, White attempted to maneuver with a small handheld gun that shot spurts of pressurized oxygen as a propellant. But within a very few minutes its fuel was exhausted.

The lasting impact of Ed White's spacewalk was solidified by his complete exuberance during the experience: He floated and whirled around the Gemini spacecraft as the Earth passed below. White was outside twice as long as Leonov had been, and it began to look as though he didn't want to come inside. The available sunlight was rapidly dwindling as he struggled back into his couch, remarking, "It's the saddest moment of my life." Journalists were fascinated by White's giddy joy, and some asked NASA physicians whether he may have been experiencing "space intoxication." In fact, his return to Gemini had taken longer than planned because he discovered that moving his body into the proper position to get inside was more difficult to execute than anyone had predicted.

Much of the back-and-forth conversation between McDivitt, White, and capsule communicator (capcom) astronaut Gus Grissom during the EVA concerned making sure the event was captured on film. White had mounted a time-lapse 16mm film camera on the spacecraft and had a small still camera attached to his handheld maneuvering unit. Inside the spacecraft, McDivitt operated a 70mm Hasselblad camera, shooting images whenever White floated into the view afforded by the command pilot's small hatch window. "Take some pictures!" Grissom reminded them. White shot a small number of photos as he drifted, but none of them conveyed any sense of what he was experiencing. In contrast, McDivitt's pictures taken through his tiny window were a revelation. They became iconic images that paid invaluable dividends and are among the most reproduced photographs of the American space program.

The photos and film of Ed White floating in space were a record of an extraordinary achievement but also an eerie reminder of the fragility of a human body suspended in the harsh environment of space. White appeared at ease while hurtling hundreds of miles an hour above the Earth, enclosed in his fragile personal spacecraft. Despite the fact that his face was obscured by a reflective gold visor, the photographs left no doubt this was an American spaceman: In almost every photograph, an American flag is prominently displayed on the upper left shoulder of his space suit, a deliberate counterpart to the CCCP on Leonov's helmet. Gemini 4 was the first time an American space suit displayed the flag. Henceforth it was worn on all U.S. space missions.

White's exuberance in space was nearly matched by Lyndon Johnson's elation with Gemini 4's success. He was particularly excited by how the news was being received in Europe and shared reports of enthralled audiences at the Paris Air Show queuing up to see screenings of newly released films of the spacewalk. He also invited the astronauts and their families for an overnight visit at the White House. Honoring them at the reception, Johnson referred to McDivitt and White as "the Christopher Columbuses of the twentieth century." In her diary, First Lady Lady Bird Johnson wrote that her husband was subtly shifting the rhetoric of the space race away from talk of a global battle for survival to one of peaceful exploration: "Men who have worked together to reach the stars are not likely to descend together into the depths of war and desolation," she wrote. It was far better that brave pilots ride "together in a spaceship to a new adventure, to discover a new world, than shoot down each other's planes [in Vietnam] as we had to last night."

NASA's own publications echoed Johnson's shift in tone. On the front cover of *A Guide to Careers in Aero-Space Technology*, widely distributed to secondary schools and colleges in the mid-1960s, the *Nina, Pinta*, and *Santa Maria* were shown in the blackness of space, accompanied with the words: "NASA 20th Century Explorer . . . into the sea of space." The Soviet challenge had not disappeared, but NASA and the journalists covering Apollo now defined the quest to the Moon as a manifestation of national will, the fulfillment of Kennedy's aspirational challenge, and an exploration to expand knowledge about the universe.

The national attention given Gemini 4 also provided an opportunity to highlight Webb's personal vision of NASA in its role as a catalyst for science education. Before being honored at the White House, McDivitt and White were feted in Ann Arbor, where they attended the dedication of the new Space Research Building at the University of Michigan. That year alone, the University of Michigan received more than 6 million dollars in NASA funding for space-related research, just one of hundreds of similar educational programs that the space agency was funding around the country. By the mid-1960s, NASA was distributing more than 50 million dollars a year to state universities for research grants and contracts. Webb intended to use the space agency's money to strengthen the academic standing of state schools that had difficulty competing for top academic talent against better-endowed prestige schools. Disbursing these state-university grants throughout the country also ensured sustained support in Congress whenever NASA's budget came up for a vote.

The NASA educational program was beginning to flourish just as Lyndon Johnson's relationship with the academic community was becoming strained. Ironically, the first anti-war "teach-in" about the Vietnam War occurred on the Ann Arbor campus a few weeks before the Space Research Building dedication; classes had been canceled and two hundred faculty members took part in a series of seminars about the colonial history in Southeast Asia and America's involvement there. Johnson was particularly peaved that his foreign policy had come under academic criticism; he had assumed the educational community shared his ideals and was among his strongest allies. But the continued campus protests against his foreign policy undermined Webb's university program, which was still too new to have evidenced the promise he had envisioned. On one occasion when fuming about his critics, Johnson complained to Webb about the "little bastards" on the campuses and demanded that Webb stop allocating NASA money to support his educational initiative. As NASA suffered continued budget cuts triggered by increased spending for the war in Southeast Asia, the university subsidies became a line item Webb was reluctantly forced to reduce.

Despite their similar and near-simultaneous political educations in Washington during the early FDR years and their shared conviction that effective big government could constructively effect social change and improve lives, Webb and Johnson were never close personally. Perhaps their temperaments were too much alike, each wanting to maintain control. Nevertheless, they worked well together, keeping out of each other's way when necessary. As the war escalated, Johnson repeatedly apologized to Webb about the sacrifices to NASA's budget, promising him that once the conflict was over, NASA would see a brighter future.

When Johnson needed his help on Capitol Hill, Webb proved a valuable White House ally, particularly when the president was looking for Southern Democratic votes to pass the Civil Rights Act and the Voting Rights Act. With so many NASA branches located in Southern states—especially after the addition of new facilities in Louisiana and Mississippi in the early 1960s—Webb was responsible for a lot of jobs below the Mason–Dixon Line. Johnson called on him as NASA administrator to exert pressure on many of the segregationist congressmen. Much of this was framed in the form of a *quid pro quo* gentleman's agreement that assured a continued influx of NASA dollars to their states. Seeing that the Southern NASA centers complied with the new federal equal-employment statutes was even more fraught, especially as the civil rights movement gained ground during the 1960s. Attorney General Robert Kennedy unleashed his fury on Webb and Johnson concerning NASA's poor record for equal-opportunity hiring in 1963, and the problem continued to plague the space agency for the remainder of the decade, though it seldom received national attention.

Additional complications arose whenever a powerful NASA-friendly politician or organization in the South attempted to use the allure of the space program to attract attention to a segregationist event or promote their agenda. NASA received a constant stream of requests for astronaut public appearances. As part of their official duties, they were expected to spend a few weeks every year bolstering public support for the space program. In accordance with the new civil rights legislation, Webb enacted strict guidelines prohibiting NASA representatives from participating with organizations that had membership restrictions.

Webb found himself in one such sticky situation when Mississippi's powerful senator John Stennis, a staunch supporter of the space program as well as a fierce segregationist, asked NASA's administrator to deliver a speech in Jackson. Only hours before he was to appear at the event in the state capital, Webb learned that the venue was a segregated Chamber of Commerce celebration where the governor was to promote a slate of politicians elected on a racist platform. The Student Nonviolent Coordinating Committee was preparing a massive protest against the event, while the local deputies were preparing to bring out police dogs and use violence against the demonstrators. Webb felt duty bound to honor his promise to Stennis, but he realized it would be a PR disaster if he did so. Shortly before the event was scheduled to take place, he called Stennis and told him that he would not appear. But he attempted to let the senator save face by giving him full permission to personally blame Webb for the embarrassing situation and telling him he could assign any motives he wished for Webb's failure to show up.

Webb was far better prepared some months later when Alabama's governor George Wallace tried to boost his national profile by inviting a contingent of journalists and state legislators on a tour of von Braun's Marshall Space Flight Center. Weeks earlier Webb had publicly criticized Wallace's segregationist policies as "backward" for Alabama. Privately, Webb regarded Wallace as a ruthless and dangerous demagogue. Concerned that Wallace intended to use NASA's recent successes in space to burnish his national profile, Webb and von Braun decided to undercut Wallace's grandstanding. They did this by first upstaging Wallace with an event that was literally earthshaking: a long-duration static test firing of an F-1 Saturn V engine. When the engine ignited, a massive plume of smoke shot away from the test stand, and there was a roar that rose to a continuous and deafening wall of sound. Those watching from the observation area were buffeted by a concussive wave of air pressure so forceful that it rippled their clothing and the men could feel the heat from the engine forced up their pant legs. This was the longest sustained test of the engine ever attempted, and after it had concluded, all present were somewhat shaken by the sheer power of what they had witnessed.

From the test area the guests were then escorted to the Marshall visitors' center, where both Webb and von Braun used the opportunity to explain to the press and the legislators why they believed the governor's segregationist policies would hinder future industrial expansion in Alabama. In his speech, von Braun even pointedly alluded to Alabama's racial history, by saying the dawning age of space technology would "belong to those who can shed the shackles of the past." Henceforth, Wallace avoided antagonizing the citizens of Huntsville, choosing other locations in the state to deliver his fiery rhetoric.

The national attention accorded von Braun's speech prompted new

The segregationist policies of Alabama's governor George Wallace prompted NASA to consider relocating their Marshall Space Flight Center from Huntsville to another state. During a Marshall press event where the power of the new Saturn V engine was demonstrated, Wernher von Braun (center) and NASA administrator James Webb (right) confronted Wallace (left) as representing the dangerously regressive ideas of the past.

focus on him as one of the nation's more unlikely civil rights advocates. Von Braun had recently appeared onscreen in an internal NASA film outlining the importance of compliance with the Civil Rights Act. He had urged Huntsville merchants to integrate the city's hotels and restaurants, explaining that their Jim Crow policies were harming local businesses because commercial aerospace contractors were unable to book rooms in the city. In a profile published shortly after the incident with Governor Wallace, *The New York Times* named von Braun "one of the most outspoken spokesmen for racial moderation in the South" and noted how Marshall was having difficulty attracting job candidates to Alabama in the wake of the negative press coverage about recent civil rights incidents in Birmingham and Selma. There was even a hint that NASA might be forced to close and relocate the Marshall Space Flight Center if the situation did not significantly improve.

As the space age coincided with the civil rights era, it was inevitable there would be periodic intersections. NBC announced it would broadcast the first American television drama series to feature a black actor in a lead role. And, once again, questions were raised in the press asking when America might recruit the first black astronaut. This time the discussion arose when *Ebony* magazine published a disturbing account of what had happened to Edward Dwight at Edwards Air Force Base two years earlier and his failure to be selected as part of NASA's Astronaut Group Three. It implied possible complicity on the part of NASA, noting that only 2.5 percent of its workforce was African American. Dwight responded with a public statement asserting that his argument was entirely with the Air Force and he didn't fault the astronaut selection process. Facing reporters at a press conference held on an Air Force airstrip in California, Dwight was asked whether he might still be selected as an astronaut. He could only reply, "I have no idea." At the Pentagon, the Air Force continued a search for other accomplished black pilots who might qualify for future astronaut training, but from NASA officials, Dwight heard nothing more.

In a bizarre coincidence, the story of Ed Dwight's treatment at Edwards coincided with the flurry of media attention given Gemini 4 and Ed White's spacewalk. And, not surprisingly, some confused astronaut candidate Ed Dwight and spacewalker astronaut Ed White. The situation was further compounded by the widely published pictures of White floating above the Earth, his face entirely obscured by his helmet's gold visor. The stunning photographs of America's first spacewalker showed a space-age Everyman, the personification of national determination and technological progress that transcended racial identity. Any viewer gazing upon the photographs could easily project an imagined identity for the anonymous person behind the visor.

In the Manned Spacecraft Center's astronaut office in Houston, Ed White was surprised to discover that among the thousands of letters he received after Gemini 4, some had been intended for Ed Dwight. Many had assumed that the first man to walk in space had also been the first minority astronaut. White put these letters aside and personally handed two boxes of correspondence to Dwight. Already moved by White's generosity and thoughtfulness, Dwight was further impressed when White, after seeing all the messages from people around the world, told him, "Now I understand why it's important for you to go into space."

All the undue media attention accorded Ed Dwight had taken a toll. Having earned a reputation as someone who didn't accept the status quo, his Air Force career was in limbo. As a result, he resigned his commission later that year. His personal life was in turmoil as well. *Ebony* had reported that he was in the midst of a divorce, a biographical detail so socially controversial in the mid-1960s that no astronaut candidate with this stigma on his record would ever receive serious consideration.

Gemini's first year concluded with a triumphant and visually dazzling accomplishment that served as a demonstration of the crewed space program's ability to remain agile and innovate even when confronted with adversity. Following the eight-day mission of Gemini 5—long enough to cover the length of time needed to travel to the Moon and back—Gemini 6 was to be the first attempt to rendezvous and dock with a target vehicle. But when the Agena target rocket was lost in an explosion over the Atlantic, the original flight plan was canceled. In its place, an audacious new idea was proposed in which Gemini 6A (a new designation for a new mission) would rendezvous with Gemini 7, already scheduled to carry out a two-week medical endurance mission. The plan would have Gemini 7 launched first, out of sequence. Then eight days later Gemini 6A would follow it into space and rendezvous with Gemini 7 using onboard radar and computers. Gemini 6A eventually got off the ground eleven days after Gemini 7 and rapidly demonstrated that with the proper equipment and precise calculations a rendezvous could be accomplished efficiently without expending excess fuel. For more than four and a half hours, Gemini 6A carefully maneuvered itself around Gemini 7 as both vehicles hurtled above the Earth at 17,000 miles per hour. At one point the vehicles came within two feet of each other.

The widely reproduced photographs taken from one Gemini spacecraft showing the astronauts' view of a second Gemini vehicle orbiting the brilliant blue planet below provided vicarious eyewitness testimony of America's progress and served to bolster James Webb's long-term rhetorical strategy. Receiving far less attention were hundreds of individual Gemini photographs looking down on the Earth, including one taken by Gemini 7 showing a hazy plume extending across East Texas, considered the earliest astronaut sighting of industrial air pollution from space. Unfortunately, over the next four years such observations became more frequent, with the astronauts occasionally requesting that NASA "call the pollution control boys" to ensure that the smoke and smog they were seeing received serious environmental attention.

After the Gemini 6A and Gemini 7 mission, the Soviet threat appeared to recede, as the United States broke two more space records. The Associated Press began distributing a "U.S.-Soviet box score," treating the space race like an international sports competition, assigning points to the country that had flown the greatest number of piloted flights, crewed flights, hours in space, astronauts in space, spacewalks, hours spent outside the spacecraft, rendezvous, and so forth. During the first year of Gemini, the Soviet Union lagged far behind, not having launched a single piloted flight.

The Soviet Union had no counterpart to James Webb overseeing the management of its space program. There was no NASA-like Soviet space agency; instead, the decentralized Soviet program was run by the military, with separate competing aerospace design groups working on space-related defense projects. This allowed small teams to improvise missions based on already established and familiar hardware and technology. But without the unified guiding vision of a central manager, the competing teams suffered from a lack of coordination.

Lacking a managing administrator, the Soviets relied on the vision of a single anonymous "chief designer" to achieve most of the early feats, such as the launch of Sputnik and the piloted Vostok flights. The chief designer was Russia's counterpart to von Braun, and his identity was a guarded government secret. However, the world learned his name in January 1966, when the death of Sergei Korolev, "the man who provided the scientific and technical leadership of the vast Soviet rocket program," was announced in the world press. It was difficult for the United States to assess the long-term impact of Korolev's death on its rival's space program. Webb continued to receive access to the CIA's top-secret intelligence briefings, and in the post-Korolev years he seldom appeared unconcerned about the Soviets' progress. But the long list of their space accomplishments since 1957 made it impossible to discount the prospect that they might surprise the world again with little advance warning.

AFTER HALF A decade of piloted missions, venturing into outer space no longer seemed as threatening as it had when Yuri Gagarin first circled the globe. The films of Alexei Leonov and Ed White floating in the void outside their spacecraft made spacewalking appear to be the adrenaline junkie's ultimate adventure sport. Astronauts brought harmonicas and paperback books into space, leading to accounts that deceptively suggested that space travel was leisurely and relaxing. But the reality of life in a Gemini spacecraft was nothing like the depictions of space travel in Hollywood films or at the World's Fair. The crew area in a Gemini capsule was roughly equivalent to the seating area in a compact car. Sleeping was difficult, especially while wearing a pressure suit, and exacerbated by the constant need to monitor hundreds of details. There was seldom any time to relax. Flight plans were filled with activities; plus there were endless mundane chores like meticulously labeling and storing daily stool samples, or, as on one Gemini mission, cleaning up a blizzard of dehydrated shrimp floating throughout the cabin after a food packet burst in the zero-gravity environment. Such details were unlikely to make imaginations soar or sell magazines. Rather, it was the authentic joy on the faces of the astronauts as they stood on the deck of a recovery ship breathing fresh air, feeling sunshine, and once again experiencing gravity that captured the public's heart.

America's ambitious plans for the second and final year of the Gemini program called for astronauts to physically dock with a target spacecraft and conduct adventurous spacewalks without the use of an umbilical cord. Astronauts would maneuver independently around the spacecraft using a self-contained rocket backpack. There was even a possibility that the final Gemini mission would fly in tandem with the first piloted Apollo flight.

For the producers at the three television networks, conveying the stories of the Gemini missions as an engaging and informative narrative continued to be a challenge. Until late 1965, the rocket launch was the only part of the mission that could be seen as it happened. This changed with Gemini 6A, when television pictures of the recovery were transmitted live from an experimental satellite earth station on the aircraft carrier USS *Wasp*. For everything between launch and recovery, the networks continued to tell the story using scale models, prepared animation, diagrams, and demonstrations. Marionettes were even used to depict a Gemini spacewalk while it took place in orbit. NBC invested one hundred thousand dollars in a full-size mock-up of the Gemini spacecraft, which was installed in Studio 8H in Rockefeller Center.

As part of their launch coverage, the network news divisions were careful not to sensationalize the potential danger, but few viewers remained unaware that as they watched a countdown they might witness a spectacular, possibly fatal, explosion at the moment it happened. Launch-day broadcasts often included a report detailing precisely how the spacecraft's escape mechanisms were designed to function in the event that an unexpected mishap endangered the astronauts' lives. Such an emergency had nearly occurred during an aborted launch of Gemini 6A. As Gemini 7 orbited above Cape Kennedy, Gemini 6A's countdown climaxed with a cloud of smoke and the abrupt shutdown of the Titan II rocket's twin engines. For a few tense seconds it was uncertain whether the stationary rocket might suddenly rise into the sky as planned or erupt in an orange fireball, offering the astronauts no choice but to fire their ejection seats, an option that carried huge risks. Luckily, the steady nerves of command pilot Wally Schirra prevailed as he gripped, but chose not to pull, the D-ring that would activate the ejection device—a hazardous but ultimately wise decision that saved the mission. The problem that caused the shutdown was swiftly remedied, and Gemini 6A rocketed toward its belated rendezvous with Gemini 7 two days later.

But the most dangerous moment during the Gemini program did not occur during a launch but while Gemini 8 was in orbit 145 miles above the Earth. When it happened, it became a two-day news story that rapidly faded from memory, but it was an event that precipitated two important decisions that were to directly impact the achievement of President Kennedy's goal by the end of the decade.

An hour and forty minutes after an Atlas missile launched an Agena docking vehicle into orbit, two rookie astronauts, commander Neil Armstrong and pilot David Scott, headed into space aboard Gemini 8. Unlike what had happened a few months earlier on Gemini 6, the Agena worked perfectly this time. When Gemini 8 was in its fourth orbit, Armstrong closed in and maneuvered near to the Agena. A member of Astronaut Group Two, Armstrong was already a famed X-15 rocket pilot, and his unusual status as one of the few civilian astronauts set him apart.

As the two spacecraft passed above Brazil, Armstrong accomplished the last remaining major goal of the Gemini program: He eased the nose of his vehicle into the docking collar on one end of the Agena and then threw a switch that latched the two vehicles together. Because Apollo would rely on the lunar-orbit-rendezvous flight plan, no landing on the Moon would be possible unless the ability to rendezvous and

Neil Armstrong's view as he moved Gemini 8 toward the Agena target vehicle above the Pacific Ocean before accomplishing the first successful docking of two spacecraft. Minutes later, Armstrong and David Scott lost control of Gemini 8 when a malfunctioning thruster sent their spacecraft into a dangerous spin, eventually forcing them to make an emergency return to Earth.

dock two freely floating spacecraft had been mastered. On his first attempt Armstrong had made it look easy, laconically describing the maneuver as "a real smoothie."

But there was nothing routine about what transpired less than an hour later. As Armstrong and Scott prepared for their first evening in space, Scott noticed that the inertial-platform display on the panel directly below his window indicated that their spacecraft and the Agena were banking at a thirty-degree angle. Armstrong rapidly straightened the attitude with the Gemini's hand controller, but once again the spacecraft began the banking motion. Assuming the problem was likely a mechanical malfunction originating in the Agena—a spacecraft with a known troubled development history—Armstrong decided to carefully undock and back away from the target vehicle, then try to regain control of his own spacecraft.

This maneuver occurred while the spacecraft was out of radio contact with the receiving stations on the ground. The first indication that Gemini 8 was experiencing any difficulty came when a NASA communications ship heard Armstrong report, "We have serious problems here. We're tumbling end over end up here. We're disengaged from the Agena."

Those listening in to the confidential audio loop heard Armstrong clearly say, "We're rolling up and we can't turn anything off."

Armstrong's actions to separate from the Agena had done nothing to solve the problem. Instead, their rotation rate had accelerated to nearly one revolution per second. Armstrong described the sensation like being inside a tumbling gyroscope. He noticed that whenever he moved his head, the rapid motion caused his vision to blur, and he thought he might lose consciousness. There was a break in the communication, and then he was heard saying, "Stuck hand control," which alarmed the flight-control team listening in.

Armstrong realized the problem had to be located somewhere in their spacecraft, most likely a malfunctioning thruster on the aft portion of the Gemini equipment module, so he chose to disengage power to all the rear thrusters. However, even with the malfunctioning thruster now out of operation, the Gemini's gyrations continued unchecked, so next he tried to stabilize the spacecraft by engaging an unused group of thruster rings located in the nose of the Gemini—devices that were intended for use exclusively during reentry into the Earth's atmosphere.

As they emerged into the daylight side of the Earth, Armstrong had begun to stabilize Gemini 8. He had been carefully coaxing the reentry thrusters for nearly twenty minutes when he reported to the Hawaii tracking station, "We're slowly getting back to the proper attitude." The mission rules dictated that once the reentry thrusters had been engaged for any purpose, the astronauts would need to return to Earth, landing in the first available recovery area. John Hodge, the flight director in Houston, decided to bring Gemini 8 home during the seventh orbit. And since they needed a daylight recovery location, the only choice was a secondary zone in the Pacific where a small contingent of ships and planes was available.

Armstrong was not thrilled about the recovery plans but kept any reservations about the decision within the spacecraft. "I'd like to argue with them about the going home," Armstrong told Scott. "But I don't know how we can. I hate to land way out in the wilderness." Weighing on his mind was how difficult spotting their spacecraft might be for the recovery forces when it was but a tiny speck floating in the middle of the ocean. He couldn't forget the story of the *Andrea Doria* and how it took rescuers a day and half to locate the large cruise ship when it began to sink off Nantucket in 1956. In comparison, their Gemini capsule was so small it would easily fit into the *Andrea Doria's* first-class swimming pool.

Gemini 8's emergency came to the attention of the three television networks just as they were to begin broadcasting their valuable prime-time entertainment programs. Ten minutes before CBS was to air the latest episode of the children's adventure series *Lost in Space,* Walter Cronkite began live coverage of the unfolding space crisis from his New York news desk. NBC followed a few minutes later, breaking into the opening minutes of *The Virginian*. Over at ABC, though the network's news division had been carefully monitoring the crisis in space, programmers in New York opted to broadcast a new episode of their hit series *Batman* instead. The camp adventure show had just introduced its latest guest villain, Catwoman, when ABC suddenly decided to break in with a brief Gemini news bulletin, the first of four that interrupted *Batman* during the next half hour, rapidly rendering the plot incomprehensible and angering millions of viewers.

Ironically, unlike their competitors at NBC and CBS, the perennially third-place network had an exclusive that evening. One of the ABC News producers had hacked NASA's communications loop and had access to the live audio from Gemini 8. But even this scoop couldn't persuade the programmers to postpone *Batman*. Eventually ABC began its live continuing coverage of Gemini 8 after *Batman* had concluded, but by then any viewers who had wanted to follow the crisis in space had already switched over to the other networks.

By 11:00 P.M. in New York, the networks concluded their live breaking coverage with news that the Gemini 8 astronauts were safe in the Pacific recovery area. It was called "one of the year's best and most suspenseful true-life dramas" by a television journalist, but viewers didn't agree. The network switchboards registered a record number of complaints from people angry about missing their favorite programs; not a single caller to NBC's switchboard expressed praise for their Gemini coverage. Overnight ratings confirmed that TV sets were gradually switched off that evening; a viewer who saw some of the Gemini 8 broadcast called it "too long, boring, and there was nothing to see." The networks lost roughly 3 million dollars in advertising revenue that night, with local affiliates forfeiting additional money as well. Most Americans appeared to prefer the fiction of *Lost in Space* to the real thing happening live.

Aboard the recovery ship *Leonard F. Mason,* Armstrong appeared depressed, refusing congratulatory handshakes. Despite achieving the world's first space rendezvous, he worried it might appear that he was responsible for wasting American taxpayer money. If somehow Gemini 8's near disaster was attributed to pilot error, Armstrong knew, this flight would probably be his last trip into space. But his concerns were quickly banished, as it became clear within hours of the return that Armstrong's quick and commanding decision-making and his skill in bringing the spacecraft's wild gyrations under control were exemplary

Hours after their near-death experience on Gemini 8, astronauts Neil Armstrong and David Scott smile at the press and crowds of onlookers at Hickam Air Force Base in Hawaii. Among those who turned out to catch a glimpse of the astronauts was four-year-old Barack Obama and his grandfather.

of the way to handle such a crisis. Armstrong had proven himself to be the ideal person to command one of the first missions to the Moon.

The Gemini 8 crew members left the *Leonard F. Mason* when it docked in Okinawa and flew to Cape Kennedy via an eighteen-hour layover in Hawaii. A scrum of photographers and the local Honolulu television stations covered Armstrong and Scott's departure for Cape Kennedy as a crowd of well-wishers looked on. In the ring of observers at Hickam Air Force Base trying to catch a glimpse of the two Gemini heroes was a four-year-old boy waving an American flag while sitting on the shoulders of his grandfather. Years later as president of the United States, Barack Obama recalled that moment as one of his earliest memories.

Despite the fortunate and safe return of Armstrong and Scott, James Webb was not pleased with how the situation had been handled within NASA. In particular, Webb was furious that while the crisis was taking place, Robert Seamans, who had recently assumed the position of deputy administrator, had addressed thousands of VIP attendees at the black-tie Robert Goddard Dinner in Washington about the ongoing events, even though he didn't have access to verifiable facts. Webb told him that if NASA had released a misleading or erroneous public statement during a developing emergency situation, the public's trust in the space agency would have been harmed irreparably.

Previous to Gemini 8, a rough plan of action had been established in case of an accident during a spaceflight, which focused primarily on investigating the cause of the incident. Less thought had been given to handling public announcements during a developing emergency situation and its immediate aftermath, and both Seamans and Webb concluded they had to be better prepared if and when the next crisis occurred. At Webb's direction, Seamans drafted a new emergency plan, recommending that NASA's top administrators use confidential lines of communication to gather accurate information before releasing any news to the public. And beyond that, any accident-review-board inquiry would be conducted as a NASA internal investigation and reported directly and solely to NASA's administrator.

During the Gemini 8 emergency, the space agency's vaunted open public-affairs program came under press scrutiny, especially when NASA withheld recordings of the astronauts' communications with Houston for nearly twenty-four hours. As was revealed when the tapes were made available, few signs of undue alarm or panic could be heard in the astronauts' voices. But NASA's desire to contain and control the narrative continued to frustrate journalists. Some reporters were convinced of the existence of a separate private communications channel between the ground and the spacecraft, used for confidential conversations that were never made available to the press.

While reviewing the Gemini 8 news coverage, NASA's public-affairs office discovered that ABC News had included exclusive information in their broadcast, leading the space agency to correctly surmise that ABC had somehow hacked NASA's internal audio-communications loop. NASA summoned the ABC News producers to a meeting at its Washington office and threatened to cut off their press access in retaliation. But the network producers feigned innocence, admitting to nothing. The situation only exacerbated the already testy relationship between NASA's Julian Scheer and ABC's ambitious science correspondent Jules Bergman, who had recently been caught venturing into restricted areas and acting as if he didn't need to follow the same rules and procedures as his colleagues.

The next shot in NASA's war with ABC came when the network's news president, Elmer Lower, complained that those currently in power in Washington had become increasingly restrictive when giving journalists access to information. In a public speech, he attacked the pervasive culture of dishonesty practiced by the government and military press officers in South Vietnam, pointing a finger specifically at the high-handed attitudes of the public-relations officers working for NASA, branding the space agency "one of the largest sacred cows in the federal establishment." Lower even reported that Jules Bergman believed that NASA was actively wiretapping phones at Cape Kennedy to prevent news leaks, a wild charge with no supporting evidence.

As part of his attack on NASA, Lower presented a case for the addition of live television on future Gemini flights. But Lower's reasons were less about granting Americans an opportunity to see their tax dollars in action than a practical business decision. Since third-place ABC News had a minuscule production budget in comparison to its two better-financed competitors, CBS and NBC, Lower knew that live television from space would immediately put the three networks on a level playing field, with all forced to broadcast the identical audio and video.

Lower charged that NASA opposed live television from space out of fear a tragedy would erode public support, arguing the public's right to know was more important than protecting a huge government agency's reputation. In fact, Julian Scheer and some others within NASA very much wanted live television included on future missions; the resistance to television cameras on American spacecraft was concentrated among engineers and astronauts.

The ABC News president's criticism of NASA reflected a gradual shift in attitudes regarding the American government that had arisen during the half decade since Kennedy's call to go to the Moon. Although the space agency's critics had garnered little national attention, there were stirrings of opposition to Apollo on Capitol Hill, in academia, and on newspaper editorial pages. As academic and intellectual criticism of Johnson's Vietnam policy grew louder, it brought with it

skepticism about the underlying wisdom of other government priorities, including renewed scrutiny of NASA and the decision to land
men on the Moon.

Nevertheless, James Webb believed that his larger vision of NASA
as a catalyst for the nation's educational, technological, and manufacturing base made it a logical ally in achieving Lyndon Johnson's Great
Society, especially if the lessons of space-age management could be
applied to other domestic challenges. So it was something of a rude
awakening when Webb was invited to address the annual National
League of Cities conference and heard mayors and city officials attack
the federal spending for Apollo as misguided when so many American
cities were in crisis and their renewal been declared a national priority.
Webb countered that neither he nor anyone else in government believed that the space program was considered a higher priority than the
needs of the cities but argued that the country should spend money
where it could effectively lead to productive change. Webb's argument
relied upon his belief that an innovative big-government program with
a clearly defined goal could, when managed creatively, serve to nurture
educational and technological spin-offs. However, his nuanced defense was somewhat undercut by NASA successfully telling its story as
one of heroic exploration rather than emphasizing the long-term benefits to the nation.

The skepticism toward government priorities that Webb encountered at the National League of Cities conference was a reflection of
the larger changing culture during the immediate post-Kennedy era.
For thirty years, from the height of the Great Depression, when unemployment exceeded 20 percent, and through the 1950s, opinion polls
indicated that Americans remained optimistic about the country's future. After the trauma of the Kennedy assassination and America's
growing involvement in Southeast Asia, such optimism began to decline. As World War II and Korean combat veterans assumed positions
of power, some influential works of art employed irony and dark humor
to satirize and call attention to the underlying absurdity and hypocrisy
of institutions, political pragmatism, and American culture in general.
Joseph Heller's *Catch-22* (1961) and Stanley Kubrick's *Dr. Strangelove*
(1964) spoke to both a generation of experienced war veterans and

their children—the first wave of the baby boom, who entered college during the early 1960s, energized by the idealism of the Kennedy era.

During the Gemini years, the American space program was largely spared the biting satire inflicted on the Pentagon and other American institutions, with one powerful exception. Tom Lehrer, an academic with connections to Harvard and MIT, moonlighted as a performer of comic songs about social and political issues. In 1965 he released a comedy LP that included a savage two-minute ballad about the career of Wernher von Braun. Confronting his Nazi past and his political expediency, the song's clever lyrics and irreverence would have been unthinkable during the Eisenhower era.

> Don't say that he's hypocritical,
> Say rather that he's apolitical.
> "Once the rockets are up, who cares where they come down?
> That's not my department!" says Wernher von Braun.

Far more than Hollywood's portrayal of von Braun in *I Aim at the Stars,* Lehrer's brief musical takedown became the lasting artistic interpretation of his life and legacy. What irked Lehrer wasn't von Braun's employment by NASA but his status as a national hero in the media, which he considered grotesque. Arthur C. Clarke provides one of the rare accounts of von Braun's reaction to Lehrer's song, recalling a party where Lehrer's LP was played and Clarke witnessed von Braun's usual good sense of humor "tested to the breaking point."

With his subversive song, Tom Lehrer focused attention on the wartime past of von Braun and the Peenemünde Germans working on Apollo—something few American journalists covering the space program in the 1960s had revisited. For them, the story of the V-2's development was old news and irrelevant to the bigger ongoing story. Von Braun's colleague from Peenemünde Kurt Debus was now Cape Kennedy's director of the Launch Operations Center. An imposing figure, with a face scarred from deep wounds sustained during his university fencing days in Darmstadt, Debus had been an active member in Heinrich Himmler's SS during the Third Reich. Journalists weren't comfortable detailing biographies such as his when writing pieces about Gemini and Apollo, even though in private they joked about the irony that a lot of former Nazis were going to put America on the Moon.

WITH CULTURAL AND political attitudes in transition, James Webb continued to strive to build support for NASA's post-Apollo future, without any encouragement from the White House. In mid-1966, congressionally imposed NASA budget cuts were scheduled to compel the first significant layoffs, with forty thousand to sixty thousand Apollo-related jobs to be eliminated later that year. Webb felt he had no choice but to go public to spur the White House and Congress to decide about NASA's future.

In a public interview, Webb declared the nation was facing a long-term space-planning crisis. Apollo was on track and on budget to land on the Moon, but few contingencies existed in the event of unforeseen delays. In addition, unless a longer-term decision was arrived at during the next year, the United States would sacrifice its position as a preeminent spacefaring nation during the post-Apollo era. By 1970, Webb said, NASA would meet its secondary goal of "establishing the managerial, industrial, and technological resources to do almost anything in space," but few in Washington appeared interested or committed to the nation's massive space investment.

If there had ever been a possibility that the year 2001 would see giant revolving space stations such as those Harry Lange envisioned for Stanley Kubrick's film or like the moon colonies of 2024 at the World's Fair's Futurama, it disappeared during the latter half of 1966. By the end of the year, the White House approved only a modest budget for an Apollo Applications program, which was to expand on the existing Apollo technology for missions in the 1970s. This initial plan called for extended lunar stays of up to two weeks in a roving mobile lab and an earth-orbiting space workshop built using existing Apollo hardware and technology. Only the space workshop would survive; renamed Skylab, it would be the home to three sets of astronauts in 1973 and 1974.

For the moment, though, the Gemini program was concluding with four more flights. Following Neil Armstrong's success, three of the Gemini missions also docked with Agena target vehicles, and two ignited the Agena's engine to boost the docked spacecraft into higher orbits. But Gemini's later spacewalkers had far less success than Ed White. Three subsequent Gemini spacewalkers all ran into severe difficulty when attempting to maneuver outside their spacecraft. They flailed and spun as they tried to position themselves in zero gravity and in the process became dangerously overheated and unable to see clearly through fogged visors. Learning how best to work effectively outside a spacecraft became the final challenge of the Gemini program.

The last Gemini astronaut scheduled to walk in space was Buzz Aldrin, a Korean War jet combat veteran with an MIT doctorate awarded for his work on orbital-rendezvous mechanics. Aldrin decided to take on the spacewalkers' problem using a novel and somewhat controversial training program. An experienced scuba diver, like Arthur Clarke and Wernher von Braun, Aldrin metaphorically likened skin diving to floating weightlessly in space. He knew that when he was diving it was counterproductive to struggle against the current. Rather, the most effective approach was to proceed slowly while accounting for the current's flow. Aldrin wanted to train for his spacewalk while wearing his space suit in a huge neutral-buoyancy water tank. Chief astronaut Alan Shepard and others rejected his idea, assuming there existed little similarity between being suspended in water and floating in space. But others at NASA arranged to try just such an experiment and rented a large pool at a boys' private school in Baltimore during off hours. Here Aldrin put in more than twelve hours of underwater training with submerged prototypes of the spacecraft, trying out new handrails, handles, foot holders, tethers, and EVA tools. He planned to pace himself outside the Gemini, taking scheduled two-minute rest periods to keep his heart rate consistently low.

When put into practice during Gemini 12, Aldrin's spacewalk was an astounding success. Subsequently, EVA training in a neutral-buoyancy water tank became standard preparation for all future spacewalkers. *The New York Times* hailed Gemini 12's near-perfect flight as a "victory over space." In recent months the *Times* had questioned the wisdom of Kennedy's lunar timetable but in this instance praised the significance of the scientific knowledge returned.

The success of the Gemini program had not only ushered in the transition to Apollo but also had fulfilled its secondary role of sustaining public interest in the American crewed space program. But despite the complications Armstrong had encountered on Gemini 8, the success of the ten Gemini missions led to a sense of complacency. No longer were the astronauts' names familiar to most Americans, nor were the personal accounts published in *Life* magazine attracting reader interest. The news divisions of the three television networks continued to interrupt regular programming for live launch coverage, but each of the later missions took on a routine sameness. Even the astronauts' personalities and profiles were strangely alike. Journalists did their best to distinguish between them with labels like: "the PhD astronaut," "the first Catholic astronaut," "the only Ivy League astronaut," "the civilian astronaut."

NASA announced its first three-man Apollo crew while Gemini was still at its midpoint. Two were experienced astronauts, whose names were already fairly well known: Gus Grissom, veteran of the second Mercury flight and commander of the first Gemini mission; and Ed White, Gemini 4's handsome and ebullient spacewalker. Joining them was Roger Chaffee, who was only thirty-one, a member of the third group chosen in 1963. While their spacecraft was still being assembled at North American Aviation's plant in Downey, California, motivational posters displayed on the factory's walls urged employee perfection. One featured a portrait of Gus Grissom and reproduced the three-word speech for which he had become famous. In 1959, Grissom had been asked to address thousands of aerospace workers at Convair's Atlas missile plant in San Diego. Grissom, who had been named a Mercury astronaut only weeks earlier, said everything he thought necessary: "Do good work."

Apollo's first flight was scheduled to launch only three months after the close of the Gemini program, maintaining the same energetic pace. The excitement surrounding the conclusion of the Gemini program led some journalists to assume that meeting the country's objectives in space before the end of the decade would be relatively easy. But Webb worried about anyone expressing overconfidence. Despite Gemini's many successes, memory of Gemini 8's close call was still immediate.

ALL I ASK . . .

DO GOOD WORK

MANNED FLIGHT AWARENESS

Astronaut Gus Grissom dominates this 1966 Manned Flight awareness incentive poster created to increase morale and encourage quality workmanship at facilities where many of the Apollo program's components were being built. The quotation alludes to a simple request Grissom asked of Convair employees in 1959 when they were assembling the Mercury Atlas missile that he hoped he would ride into space.

When Lyndon Johnson met with the Gemini 12 crew in front of cameras at the LBJ Ranch following their return, James Webb was standing at his side. Johnson spoke in a hoarse voice, the result of minor throat surgery the previous week. He shortened his planned talk but heeded Webb's suggestion that he insert a few words of caution when looking forward to Apollo. "The months ahead will not be easy as we reach toward the Moon," Johnson said, prompting some journalists to wonder if Kennedy's deadline was being reconsidered. But Johnson said he was merely calling attention to Apollo's many untested systems, which were far more complex than anything on Gemini.

Plans for the first Apollo flight were on schedule at Cape Kennedy when President Johnson and James Webb appeared together again, eight weeks later. The venue was the East Room of the White House; the occasion, the signing of the Outer Space Treaty. The treaty established the legal foundation for all subsequent international space law, specifically placing a prohibition on any nuclear weapons in space, on the military use of any celestial body, or on claims of sovereignty over any celestial resource. Henceforth, outer space would be used for peaceful purposes only.

Fortuitously, the treaty signing coincided with a Washington gathering of the Apollo Executives Group, composed of elite decision makers from NASA and the leading aerospace contractors. Immediately after the White House signing, Webb, von Braun, and other NASA center directors, including Houston's Robert Gilruth and the Kennedy Space Center's director Kurt Debus, as well as the heads of North American Aviation, McDonnell Aircraft, and Grumman, gathered for cocktails at the International Club, a short distance from the White House.

At roughly seven-thirty that evening, the relaxed mood began to change. Lee Atwood, CEO of North American Aviation, and the Cape's Kurt Debus received paged messages asking them to take urgent phone calls. Next was Robert Gilruth. When Atwood returned, he looked grave and took Webb aside. Webb next rushed to locate an available phone and immediately placed a call to the White House switchboard, where he was connected to one of Lyndon Johnson's secretaries. Webb dictated a short message, which was typed on a small piece of White House stationery, with a notation of the time. The note was then folded in half.

The message was taken to the president's private quarters and passed to Lyndon Johnson, who was attending a party in honor of his retiring secretary of commerce. Johnson unfolded the note as the departing secretary was making a toast. As he read it, Johnson recalled, "The shock hit me like a physical blow."

As everyone looked to the president, they noticed the sudden change in his demeanor. In a somber voice, Johnson then read the entire message out loud:

"James Webb just reported that the first Apollo crew was under test at Cape Kennedy and a fire broke out in their capsule and all three were killed. He does not know whether it was the primary crew or backup crew but believes it was the primary crew of Grissom, White, and Chaffee."

CHAPTER FIVE

EARTHRISE
(1967–1968)

Once each month, the small cohort of Florida-based journalists gathered for an informal evening of drinks, dinner, jokes, stories, and rumors. They shared a common beat: the American space program. Informally dubbed the Better Health and Sunshine Club, the gatherings would often begin in the afternoon with a round of golf. This Friday the dining room at the Rockledge Country Club, a few miles inland from Cape Kennedy, was the chosen venue.

The twenty predominantly male journalists and press officers who had congregated at the large table were in good spirits. Their recently published stories reflected a general sense of optimism. Not only did it appear likely that President Kennedy's lunar goal would be achieved before the deadline, but it was widely believed that the United States had established its lead in the space race. After ten piloted Gemini missions, the first Apollo flight, with Grissom, White, and Chaffee, was scheduled to launch in about a month. The Soviets hadn't placed a single cosmonaut in orbit for nearly two years.

As dinner was being served, the bartender approached NASA's veteran public-affairs officer Jack King, the man whose voice had provided the audio countdowns for nearly every launch from Cape Kennedy. He had an urgent phone call. He assumed it was probably another call concerning a NASA manager who had been caught speeding or arrested for DUI.

King picked up the phone at the bar, about twenty feet away from the table. After observing him say a chipper "hello," those looking on from the table saw his entire body grow tense and his facial expression become stern. He hung up the phone and stopped at the table to say, "I gotta go. Something's happened at the Cape." King would only add, "It's urgent. You'll hear from me later tonight."

Some of the reporters at the table were still on deadline and headed off to find one of the country club's few available public telephones, to check in with their contacts at the Cape. Within minutes a local reporter from the *Cocoa Tribune* learned that the prime crew of the first Apollo mission had been killed in a launchpad fire. As word spread throughout the dining room, all the journalists departed and drove back to their offices. Few would get any rest during the next week.

In the small El Lago suburb of Houston, astronaut Bill Anders had been spending the early evening working in his yard when he was called inside to take a phone call. Fellow astronaut Alan Bean regretfully told him the tragic news from the Cape and informed him that he had been assigned the solemn task of officially notifying Ed White's wife, Pat, of her husband's death. It was a duty that each astronaut knew he might have to perform someday. Anders changed his clothes and arrived at the Whites' house ten minutes later. The Whites lived next door to Neil Armstrong, who was away in Washington for the signing of the Outer Space Treaty. When Anders arrived, Pat White was on the porch, talking with Armstrong's wife, Jan. Had Neil Armstrong called her from Washington and asked her to be there as well? Anders wasn't sure, but as he walked up to the house and looked into their faces, he received an impression that Jan Armstrong had a sense of the dreaded news he was about to relay.

NASA's deputy administrator Robert Seamans was in his office at NASA headquarters, after being called away from a private dinner at his home. He had drawn up the agency's plan of action for handling an ongoing emergency, instituted after the crisis during Gemini 8. Only secure private phone lines were to be used. It was the beginning of a very long night. First, he and NASA's director of the Office of Manned Space Flight, George Mueller, compiled a list of candidates to sit on an accident review board. Following the plan, the investigation would be conducted swiftly and contained within the space agency. Both Seamans and James Webb feared that an independent board of review could take months or longer and endanger the future of the Apollo program itself. With Mueller he looked for people with no previous management role in Apollo, people who could remain objective. An astronaut would need to be included on the board, sitting in what was likely to be its most high-profile public position. In consultation with Deke Slayton, they chose Frank Borman, the veteran commander of Gemini 7. Borman's mature and no-nonsense approach to problem-solving made him an obvious candidate. Following Gemini 7, he had advanced to the Apollo program and was one of the few astronauts already well acquainted with the spacecraft. Heading the nine-member board would be Dr. Floyd Thompson, director of NASA's Langley Research Center, which hadn't been closely involved in Apollo planning at that time.

In a nearby office, public-affairs chief Julian Scheer refused to allow any NASA spokesperson to confirm news of the tragedy until all three wives had been notified. Locating and informing Gus Grissom's wife, Betty, was taking longer than expected.

In the midst of the tense two hours while the media held off breaking the news, Seamans was on a call with defense secretary Robert McNamara. Suddenly, in mid-sentence, Seamans heard a telephone operator cut into the line. She had been ordered to interrupt the call immediately by NBC reporter Peter Hackes, who had informed her it was a matter of national emergency. "The word is out! The country is almost in a panic kind of frame of mind," Hackes told him. "You've got to go on TV at eleven o'clock and reassure them."

Seamans refused. He told Hackes, "I can't do that. . . . I don't know all the facts." Nevertheless, NBC was on the air a few minutes later with a thirty-minute special report. Assured that the three widows had been informed, NASA spokesman Paul Haney confirmed the death of the astronauts in a brief statement. Initial news accounts said all three had died instantaneously, an erroneous detail that was widely reported until a journalist revealed two days later that the astronauts had lived for at least a minute after the first sign of fire and had, in fact, died as a result of smoke inhalation.

Within a few hours of the tragedy, a television report was already speculating that the fire started with a spark in the communications wiring, located near Grissom's couch on the left side of the spacecraft. In a special report, NBC's Hackes explained that the 100 percent oxygen atmosphere in the Apollo spacecraft would have made the fire burn especially quickly, noting that the Russians had chosen not to use a 100 percent oxygen environment in their vehicles. The network's Capitol Hill correspondent, Hackes also offered insight about changing congressional attitudes regarding the expense of the space program. In light of competing demands to fund the war in Vietnam and Great Society programs, he predicted new calls to further reduce NASA's budget.

Far less objective was CBS's Walter Cronkite. "This is a time for great sadness—national sadness and certainly the personal sadness of the people in the space program. But it's also a time for courage. And if that sounds trite, I'll change the word[s] to *guts*. . . . These guys who went into it knew it was a test program . . . [that] was bound to claim its victims. . . . It should not be a cause for our turning back or having any question of faltering in our progress forward toward the landing on the Moon. . . . It shouldn't in any way damage our national resolve to press on with the program for which these men gave their lives."

FRANK BORMAN'S SLEEK T-38 Talon roared off Houston's Ellington Field runway shortly after sunrise the next morning. He was piloting the twin-jet, high-altitude supersonic trainer directly across the Gulf of Mexico toward Cape Kennedy rather than following the coastline, due to the urgency of the situation. As the sky on the eastern horizon began to lighten, he pushed the engines to their limit to hasten his arrival. Alone with his thoughts in the sheltered cockpit, the thirty-eight-year-old Air Force colonel reflected on the events of the past twelve hours.

The previous evening, a Texas Ranger had unexpectedly arrived at a remote lakeside cottage where he, his wife, Susan, and his two boys had just unpacked for what they hoped would be a leisurely weekend away from their Houston home. Close friends had invited Borman and

Astronaut Frank Borman climbs into the cockpit of a NASA T-38 Talon. The space agency maintained a small fleet of the two-engine jet trainers for the astronauts to use whenever traveling to facilities around the country. Extensive hours in the T-38 prior to launch helped the astronauts hone their flying skills and physically prepare for weightlessness during space flight.

his family as weekend guests, but the astronaut had left no word of his plans with the Manned Spacecraft Center. Nevertheless, the Texas Rangers had somehow tracked him down and told him to call Houston. Borman had been instructed to drive home immediately and on Saturday morning fly a T-38 to the Cape to join other members of the accident review board investigating the causes of the fire. Leaving the boys with their hosts, he and Susan returned to Houston and headed first to the Whites' home. Nearly all the astronauts lived close to one another in El Lago. When they arrived, Bill Anders, his wife, Valerie, and Jan Armstrong were consoling Pat White.

Since moving to Houston in 1962, Susan and Pat had become close friends. Susan remained with Pat while Frank returned to their home nearby to catch a few hours of sleep before his morning flight to the Cape. But as much as she tried to be there for her friend, Susan couldn't help but imagine herself in the same situation. In the Whites' living room, the astronaut families gathered and talked until three or four in the morning. Some later remembered it as a rare moment that brought forth a frank and heartfelt discussion about the ultimate meaning of death.

Borman had been two classes ahead of Ed White at West Point, and by the mid-1950s both were part of the Air Force's global military presence during the Cold War: White flew F-86 Sabre jets in Germany; Borman was with a fighter-bomber squadron in the Philippines. Not long after, both graduated from the Edwards test-pilot school, although in different classes, and in 1962, Borman and White were among nine men chosen as members of NASA's second group of astronauts, "the New Nine."

As he flew his T-38 toward the Cape, Borman heard the voices of anonymous air-traffic controllers offering him brief messages of emotional support. The news of the Apollo fire was all over radio and television. The T-38's NASA call sign was all that was needed to explain the reason for its unusual early Saturday-morning flight plan.

Notice of a sudden accidental death was something military jet pilots and their wives had learned to accept as a part of their lives. In the Air Force, Borman had seen his share of aircraft fatalities and had served on accident review boards. It was no different in the space program. In a little more than two years, three of the twenty-eight Gemini astronauts had been killed in T-38 crashes. But this wasn't another aircraft accident. It was a sudden fire erupting inside a sealed spacecraft. And it happened while it was on the ground and during a test no one believed was especially dangerous.

The second-oldest of the New Nine, Borman had emerged as one of its leaders. He avoided unnecessary chitchat, more often speaking with a calm but forceful authority that seldom prompted a rebuttal. A single ice-cold stare from his penetrating eyes could silence a room in seconds. After heading the two-week Gemini 7 mission with Jim

Lovell, Borman had been chosen to command one of the early Apollo crews, expected to test the lunar module in earth orbit. Now, as a result of the tragedy on Pad 34, that mission and the future of the entire Apollo program were in doubt.

On the Apollo review board, he would represent every future astronaut who would enter that spacecraft, lie on its couches, and fly it into space. He wanted to make sure no astronaut did so without having full confidence it was the finest and safest spacecraft that could be built.

As he brought his T-38 in for a landing on the Cape Kennedy Air Force Station Skid Strip, Borman could see the first Apollo crew's Saturn 1B partly shrouded by the red service structure. The entire area surrounding Pad 34 had been locked down for the past few hours. After checking in to a motel and renting a car, Borman joined NASA officials and members of the review board and headed to Pad 34. There they took the elevator to the white room surrounding the Apollo command module. As they grew closer, they could see discarded fire extinguishers and cables strewn on the floor. Twenty-six workmen had fought the blaze; two had been hospitalized.

The outside of the spacecraft was partially blackened with soot, and through the open hatchway Borman could see the center couch where Ed White had been lying a few hours earlier. In the air was a strong odor of burned paper and foam material. A layer of dark-gray ash covered everything on the left side of the spacecraft, while on the right there were flight manuals and other things that appeared slightly browned but otherwise untouched. In the center, prominently visible, were twin oxygen hoses that had been attached to White's space suit. Both showed signs of having been severed with a blade as White's body was removed.

JAMES WEBB AND NASA were now facing a crisis that he and Robert Seamans had considered only an imagined possibility when they drew up the emergency plan following Gemini 7. Members of the review board were already conferring at the Cape when Webb traveled to the

The charred interior of the Apollo Command Module photographed shortly after the fire on Pad 34 that claimed the lives of astronauts Gus Grissom, Ed White, and Roger Chaffee on January 27, 1967.

White House to brief President Johnson. He was ushered into the president's private quarters, where Johnson, on this late Saturday morning, remained dressed in pajamas. The president's science adviser was already urging him to appoint an independent presidential commission, something Webb wanted to avoid. NASA's adversaries could use an open-ended investigation to weaken the agency, justify additional budget cuts, and, as a consequence, delay the lunar landing.

Webb outlined two options for the president: a NASA investigation or an independent commission. While doing so, Webb cautioned the president that his enemies could use an independent investigation to unearth details about a Johnson aide who had already been implicated in an influence-peddling deal with a lobbyist for North American Aviation. The 1968 presidential election was on the horizon, and Johnson would certainly want to avoid any reminders of past scandals, especially if they might be peripherally connected to the death of three national heroes.

"I want you to handle the investigation," Johnson said. Webb agreed, not mentioning the NASA board he and Seamans had put together the previous evening. He pointed out that Johnson could always change his mind, only requesting that he be told in advance if the president did so.

"I want you to do it," Johnson said again, and extended his hand. This was the only time Webb and the president had sealed an agreement with a handshake. Johnson had given Webb precisely what he wanted. The space agency would oversee and contain the entire accident-review process.

However, within a few hours NASA had already run into a public-relations disaster. Most of the details NASA had provided the press were correct, but the assertion that the astronauts had died instantaneously came into question when a reporter broke a story that audiotapes that captured the astronauts reporting the fire and calling out, "Get us out of here!" When reporters questioned Julian Scheer about the tapes, he denied their existence, relying on mistaken information he had been given by NASA officials in Houston and the Cape. A day later, after learning the recordings did indeed exist, Scheer issued a correction. But NASA's reputation for honesty was badly damaged, and it took more than a year for the agency to repair its relationship with the press corps.

Journalists covering the space program were already used to NASA officials keeping engineering and safety concerns from the public. Past problems had been minimized, as it was assumed that if such information were more widely known, morale and public support could erode. But after the horrific accident, reporters who had been accommodating in the past began to aggressively push back against official stonewalling. They questioned the wisdom of the moon program, its expense, and its objectives. They sought to uncover whether the three astronauts' deaths may have been preventable.

Quietly, James Webb instituted his own investigation as well. He asked NASA's lawyers to conduct a separate internal review to determine if any managerial lapses might have contributed to the fire. If there was a flaw in his system of managerial oversight, he would discover it. In the meantime, he reasoned that the only way to sustain the Apollo program and guard the space agency against its critics would be to act as a human lighting rod. He would personally absorb the blame and responsibility and accept the consequences. Webb undoubtedly remembered Truman's famous motto from when he had served as the president's director of the Bureau of the Budget: "The buck stops here!"

BORMAN AND THE other members of the Apollo review board started by impounding documents and data about the accident and collecting eyewitness accounts. Early meetings focused on reviewing all the existing technical circumstances at the time of the fire, and within days it became evident that no single cause could be isolated. Once the Apollo 1 command module was lowered from its position atop the Saturn 1B and removed from Pad 34, the review board began its physical inspection. Borman entered the interior, and his first task was to lie on the astronauts' couches and dictate the position of every switch on the control panels to accurately document their state at the moment of the inferno. A second, identical spacecraft was flown in from the North American plant, and both were carefully disassembled one piece at a time, every item examined in detail.

What the review board discovered was disturbing. Although the precise origin of the electrical spark that had caused the fire was impossible to determine, the board concluded the accident had been the result of a combination of bad decisions: the choice of a pressurized cabin with a 100 percent oxygen atmosphere; a hatch design that required at least ninety seconds to unlatch and open from the inside; the inclusion of combustible materials inside the spacecraft; vulnerable electrical wiring; and the use of a combustible and corrosive coolant. The inspection also revealed numerous instances of poor installation, design, and workmanship. When the spacecraft was taken apart, the board even discovered a forgotten socket wrench left inside the spacecraft by a North American workman months before.

Most shocking, however, was the conclusion that the fire had been entirely preventable. In addition to a series of bad engineering choices, the fire may have been exacerbated by the strained relationship between NASA personnel and North American Aviation's management. For all of Webb's administrative checks and balances, a caustic internal NASA memo—which came to be known informally as "the Phillips Report"—documented deficiencies in North American's workmanship and management more than a year before the accident. But the memo had been kept from Webb's eyes. A second safety-and-reliability memo, authored by an outside contractor only four months before the tragedy, also warned of a possible internal fire danger. The head of the Apollo Spacecraft Program Office had deemed the memo not worth pursuing, assuming any risk of a fire was remote and that instituting preventive action would cause delays and harm working relationships.

It was a moment of reckoning. NASA and its contractors experienced a sudden crisis of confidence. The space agency's reputation as the gold standard for managerial and technical brilliance was tarnished. Engineers who had placed their faith in raw data and systems analysis were shaken to realize the fatal implications of things they had overlooked or discounted. Armstrong and Scott had nearly perished on Gemini 8 due to a poorly designed electrical circuit in a thruster. Now a series of seemingly banal mistakes had killed three brave men during a routine test. As the investigation progressed, Borman watched the fire claim additional victims. A few colleagues already overworked and emotionally invested in the country's quest for the Moon became so overwrought that they suffered emotional breakdowns, incapacitating exhaustion, and episodes of mania. One consultant was literally carried out of an accident-review-board meeting in a straitjacket.

Six months earlier, James Webb had confronted Lyndon Johnson on the front page of *The New York Times* about the nation's future space planning. Now the immediate future of both Apollo and NASA was in doubt. Webb sensed a growing public skepticism toward technology. For the time being, he cautioned, NASA should avoid any language that implied "we are the apostles of a new ideology—as a new semi-religion of technology." Rather, in the wake of the tragedy, Webb believed NASA should project an image of humble workmen struggling to overcome a setback.

Webb was prepared to assume the role of the man in the crosshairs when he appeared on Capitol Hill. Congressmen and senators who for years had avoided criticizing the massive space program, for fear of appearing unpatriotic, now responded to the country's grief by putting NASA's management on the hot seat in public hearings. But it was mostly political theater, with elected officials using the media for personal publicity. Some questioned how an accident inquest overseen by the space agency could possibly remain impartial. Others wasted time asking prosecutorial questions about technical matters they didn't understand.

The Senate hearing's most dramatic moment came when Minnesota's freshman Democratic senator, Walter Mondale, ambushed Webb with questions about the 1965 internal Phillips Report critical of North American Aviation. Mondale had obtained a copy leaked to ABC's science correspondent Jules Bergman. Up to this point Webb had been unaware of the document's existence, and when responding to Mondale he appeared defensive and unprepared. The experience damaged Webb's reputation in the press and irreparably affected his relationship with members of NASA management, upon whom he had previously relied without question. Ironically, the Phillips Report had little bearing on the specific causes of the fire, though it did reveal the long-existing tension between NASA and one of its prime contractors.

Fortunately for Webb and NASA, another figure at the center of the Apollo fire investigation soon garnered far more attention from the television news cameras than the middle-aged Washington bureaucrat in charge of the space agency. With a name, face, and biography already familiar to the public, Frank Borman was the one member of the investigation review board whose every word prompted interest.

Borman arrived in Washington already anointed an astronaut hero. But on Capitol Hill and in the press, Borman was valued as a direct, articulate witness unafraid to speak his mind. Webb considered Borman the ideal public face of NASA: a humble, serious, hardworking, and patriotic American committed to fulfilling Kennedy's mandate. Since joining NASA in 1962, Borman had thought of himself as a soldier in the Cold War, assigned to a new and exploratory field of combat. As they headed to the capitol in the back seat of Webb's signature chauffeured black Checker, Webb told Borman, "I don't want you doing anything to try to protect me or NASA. The American people have a right to know exactly the unvarnished truth, and I want you to tell them."

Borman detailed what he and the board had discovered. Despite an established system of checks and balances, poor design decisions made by North American had been allowed to slip past NASA's management safeguards. However, Borman's faith in the system was not eroded; rather, he was convinced that once the design weaknesses were corrected, a far safer spacecraft would result. And in his most memorable assertion, Borman stated publicly that he would gladly fly the Apollo spacecraft with confidence after all the recommended changes had been made.

Before the news cameras, Borman projected a mature gravitas that differed from that of other high-profile astronauts. He was neither as ingratiating as John Glenn nor as free-spirited as some of the other Mercury astronauts. When responding to questions, Borman trained his blue eyes on congressmen with a gaze that could be both intimidating and disarmingly transparent. There was never any doubt about the honesty and integrity behind Borman's words. And, true to Webb's advice, Borman never hesitated to offer criticism where he thought it was deserved, calling out both NASA and North American for their errors.

Nothing said on Capitol Hill ultimately proved as persuasive for the future of Apollo as a few words Borman delivered bluntly during a special hearing of the House's Committee on Science, Space, and Technology. After asserting unhesitatingly his eagerness to fly the redesigned Apollo, Borman stated, "You are asking us if we have confidence in the spacecraft management, our own training, and . . . confidence in our leaders. I am embarrassed because it appears to be a party line. The response we have given is the truth. We are servants of the Congress and the people. We are trying to tell you that we are confident in our management, and in our engineering, and in ourselves." He then addressed the House members from the speakers' table: "I think the question is really, are you confident in us?"

Borman had put the House committee on the spot. Previous to that moment, the committee's chairman had believed the members were prepared to vote to delay the lunar program, perhaps indefinitely. But suddenly everything in the room had changed.

A West Virginia representative who a few days earlier had called for sweeping changes to NASA's management now asked for a moment to speak. "Mr. Chairman, I think we ought to end these hearings just as fast as possible and get on with the space program."

"Amen" was the chairman's only response. After a second of silence, the hearing room erupted in spontaneous applause.

Borman followed up his appearances on Capitol Hill with interviews on the television networks' Sunday political-discussion programs. Some even wondered if Borman might have a political future in Washington. When he returned to Houston, he was asked by the Manned Spacecraft Center's director Robert Gilruth to temporarily step away from the astronaut team once again. His work on the review board had proven so valuable that NASA wanted him to oversee the Apollo spacecraft redesign, working with North American in California.

NASA was now demanding zero-defects perfection from North American, and Borman would be there to make sure they complied. The redesigned spacecraft would include an improved hatch that could be opened from the inside in five seconds. All combustible materials would be eliminated from the cabin, and there would now be an additional emergency oxygen supply system. The spacecraft's atmospheric environment was changed to transition from a pressurized oxygen/nitrogen mixture while on the launchpad to a pure-oxygen atmosphere when entering space.

Working with the engineers and designers at North American would take many weeks, but neither Borman nor Gilruth imagined it would consume nearly an entire year.

HIS PUBLIC REPUTATION damaged, Webb remained as NASA's administrator despite newspaper editorials calling for his removal. He had instituted major management changes within NASA and privately de-

manded North American do likewise. But North American's CEO refused to comply. Webb had refrained from openly criticizing them
during his Capitol Hill testimony, but he leaked word that NASA was
considering transferring the Apollo contract to one of a number of
other aerospace vendors. North American's chief knew, however, that
if another contractor took over the spacecraft contract at this late date,
it would be impossible to achieve the Moon landing before the end of
the decade—in fact, it might even mean the Soviets would get there
first. After weighing the situation, though, North American's CEO decided it was wiser to work with NASA than to risk calling Webb's bluff.
He backed down and reluctantly replaced his head of the Apollo project. Webb had once again prevailed.

The news of the changes at North American made newspaper front
pages, and within days the calls for Webb to resign disappeared. But
the space agency still had a major image problem with the press and
public. During the hearings on Capitol Hill, cynical journalists began
referring to NASA's acronym as "Never A Straight Answer." Every few
months *The New York Times* published editorials questioning the purpose of the entire Apollo program and Kennedy's deadline. A NASA
official remembers jaded journalists of the time acting as if NASA was
nothing more than a big national façade overseen by a "bunch of bums"
who "didn't know what the hell [they] were doing."

By mid-1967, the percentage of Americans convinced that putting a
man on the Moon was worth the expense had fallen to 43 percent. Few
at NASA or in Washington were attempting to revive President Kennedy's aspirational rhetoric from his Rice University address five years
earlier. Yet it continued to reverberate in the popular culture. Filmed
shortly after the Apollo fire, a scene from an episode of NBC's *Star
Trek* is a case in point. Captain Kirk wants to persuade the starship
Enterprise's doubtful Dr. McCoy to consent to a dangerous experiment. Balancing potential perils with possible benefits, Kirk alludes to
the bold idealism of the United States in the late 1960s:

"Do you wish that the first Apollo mission hadn't reached the Moon,
or that we hadn't gone on to Mars and then to the nearest star? That's
like saying you wish that you still operated with scalpels and sewed
your patients up with catgut like your great-great-great-great-grandfather

used to. . . . Risk. Risk is our business. That's what the starship is all about. That's why we're aboard her."

Even if a majority of Americans no longer supported the Apollo program, prime-time entertainment had no reluctance to do so. *Star Trek* justified it as a decisive moment in the inevitable march of human progress—rhetorically framed through the historical perspective of a fictional hero three hundred years hence.

NASA was about to undertake the most complicated and dangerous series of piloted missions ever attempted in the history of spaceflight. Its image weakened, it was unlikely to survive another disaster.

A YEAR HAD passed since the last live television broadcast of a rocket launch from Cape Kennedy. Not that long ago, Titan missiles carried a Gemini crew into space nearly every other month. But the Apollo 1 fire had brought about an abrupt pause. The maiden test launch of the Saturn V moon rocket promised to be like nothing anyone had seen before. More than 250 journalists were present at the Cape's new press site near the Vehicle Assembly Building, where the Saturn V had been put together. They were there to witness the most powerful rocket ever constructed and to find out whether the United States could put the recent tragedy aside and move forward.

On this early morning launch in November, the Apollo 4 command module was unoccupied. And no journalist would be allowed within three miles of the launch site. It was impossible to reliably predict the extent of the damage should the 363-foot-tall rocket explode on liftoff.

Wernher von Braun's team at the Marshall Space Flight Center had begun working on the Saturn V in January 1960, the same month that John Kennedy announced his candidacy for president of the United States. Had a normal development schedule been followed—with each stage of the rocket tested one at a time in a series of progressive launches—the Saturn V would have taken many more years before it was ready to carry a crew. But George Mueller had persuaded von Braun to test all three live stages together during this first test flight. It was an audacious and risky option that, if successful, would save millions of dollars and eliminate months of additional testing.

At 7:00 A.M., television audiences tuning in for the *Today* show and *CBS Morning News* got a dramatic look into the future as the Saturn began to gradually rise from the launchpad. At that moment the five F-1 engines were consuming fifteen tons of liquid oxygen and kerosene per second, producing energy equal to the combined power of eighty-five Hoover Dams.

Sitting in the press site, some of the journalists noticed that the corrugated-metal roofing covering the outdoor bleachers was beginning to vibrate, and reporters could feel the force of a concussive shock wave beating against their faces. While describing the launch on CBS, Walter Cronkite and a producer noticed the large window in their broadcast booth was starting to vibrate, and both attempted to keep it from dislodging. With the audio roar of the Saturn's engines crackling in the background, Cronkite yelled, "This big glass window is shaking! We're holding it in with our hands! Look at that rocket go!" In the nearby Launch Control Center, NASA engineers seated at their consoles watched as plaster dust from the ceiling fell on their workstations.

The launch was a magnificent physical display of harnessed chemical power that left those present in awe. The events that swiftly followed were equally spectacular, though they occurred absent any eyewitnesses. Above the Earth's atmosphere, each of the individual stages of the Saturn V operated precisely as expected. Mueller's gamble had paid off. In orbit the Saturn's third stage, the S-IVB, was reignited, an operation necessary when sending a future Apollo spacecraft on a trajectory to the Moon. Then the Apollo service module's engine was successfully test-fired as well. And finally the command module separated from the service module and reentered the Earth's atmosphere at thirty-six thousand feet per second, simulating conditions a three-person crew would encounter when returning from a lunar mission.

On its first test, Wernher von Braun's Saturn V had proven itself a spectacular feat of engineering, and the United States had solidified its preeminent position in space. Far less certain was how far the Soviet program had fallen behind the United States. Western journalists had only limited information available to them. Not long after the Apollo fire, a Russian had entered space for the first time in two years. However, as with Apollo 1, this mission ended with a national tragedy. The first test flight of Soyuz, a large, sophisticated new spacecraft, was cut short less than twenty-four hours after launch, and during its return to Earth the spacecraft's parachutes failed to deploy properly. Vladimir Komarov, a veteran of the 1964 three-man Voskhod 1 mission, was killed on impact, becoming the first human to die during a space mission.

It remained unclear whether the Soviet Union was indeed preparing to land a cosmonaut on the Moon or merely attempting to deceive intelligence observers. However, James Webb was one of a select few given access to CIA National Intelligence Estimates about the Soviet

In 1967, the United States and the Soviet Union were readying their huge moon rockets for test launches. The Apollo 4 Saturn V (left) was successfully sent into space in November of that year. Russia had hoped the first flight of the N-1 (right) would coincide with the fiftieth anniversary of the October Revolution in 1967, but the initial attempt to launch the thirty-engine giant didn't take place until February 1969.

piloted space program. Webb had read an NIE briefing two years earlier that concluded with near certainty that the Soviets were committed to a piloted lunar-landing program. It forecast a crewed circumlunar
mission for the fall of 1967, timed to commemorate both the fiftieth
anniversary of the Russian Revolution and the tenth anniversary of
Sputnik.

By 1967, an updated NIE included information about two new Soviet rockets. The Proton, American analysts believed, could be config-
ured to take cosmonauts on a modest voyage around the Moon or used
to place a small space station in orbit. Also under development was a
larger, heavier-lifting rocket, somewhat comparable to von Braun's Saturn V. The CIA assumed this was intended for a piloted lunar-landing
mission, but the report hedged when addressing whether the Russians
were in a space race with the United States. The Russian lunar program was "probably not intended to be competitive with the Apollo
program," it said, but might be accelerated in hopes of getting to the
Moon first. Should they pursue that path, the 1967 CIA report continued, the earliest possible Soviet lunar-landing attempt wouldn't occur
until mid-to-late 1969.

When testifying on Capitol Hill, Webb seldom failed to mention his
concerns about the Russians' progress and would sometimes crypti-
cally refer to reports of a massive Soviet booster nearing flight readiness. Cynics referred to it as "Webb's Giant," a bit of fear-inducing
fabulation, much like threats of a new and powerful Soviet submarine
that surfaced in House testimony whenever the review of the Navy's
budget was under way. But half a century later, declassified CIA intelligence documents and spy-satellite photographs indicate that Webb's
facts about the large Soviet rocket—now known as the N-1—were
largely accurate. No conjured fabulations were necessary. On either
side of the globe, teams were assembling large moon rockets with the
hope that their work would culminate with the first human visit to an
alien world.

FRANK BORMAN HAD been working at the North American plant in
Downey, California, on a Saturday morning in August 1968, when he

received word that he was to return immediately to the Manned Space-craft Center. After close to a year rethinking every detail of the Apollo command module with its prime contractor—now renamed North American Rockwell, following a merger with automotive-component manufacturer Rockwell-Standard—Borman had returned to his astronaut duties. He was to command one of the first Apollo missions, currently slated to test the new lunar module in earth orbit. Also working on the command module that morning were the other two crew members who would join him: Bill Anders and Jim Lovell.

Anders was one of the last members of the third astronaut group to garner a prime crew assignment. He had served on the backup crew of Gemini 11 with Neil Armstrong and would be the lunar-module pilot on Borman's Apollo mission. Anders and Armstrong were among the astronaut corps' lunar-module specialists, each logging multiple flights on the Lunar Landing Research Vehicle, a finicky and very dangerous spider-like training apparatus. Jim Lovell had just been named to the crew the previous month, replacing Michael Collins, who had to undergo surgery. Lovell had spent two weeks in space with Borman during Gemini 7 and had commanded Gemini 12 with Buzz Aldrin.

Borman had been given no details of why he was needed back in Houston, but both Lovell and Anders thought it likely that something had arisen affecting their Apollo mission. They were slated to be among the first to fly the Saturn V, but the second test flight of von Braun's giant four months earlier hadn't been as successful as Apollo 4. The launch of Apollo 6 in early April 1968 had been plagued by a series of complications. Severe oscillations stemming from an unstable combustion in one of the first stage's five big engines had damaged the fuel lines and engines in the second and third stages. Additionally, the lunar module was also experiencing a series of development delays, and it appeared that it would not be ready for its first piloted test flight until the following year.

Borman flew his T-38 from California to Houston, and by the early afternoon he was sitting in Deke Slayton's office. Slayton instructed him to close the door. He told Borman that there were CIA reports indicating a massive rocket was being readied at the Soviets' Baikonur

Cosmodrome. The giant Russian moon rocket that James Webb had spoken of on Capitol Hill was indeed a reality, and intelligence analysts believed there was a good possibility that the Soviets might attempt a piloted lunar-flyby mission before the end of the year.

With the lunar module experiencing delays, Slayton asked Borman to consider a radical and risky idea that had been proposed by George Low, the head of Apollo's Spacecraft Program Office. Rather than waiting for a flight-ready lunar module, the United States could counter a potential Soviet challenge by sending Borman's crew to the Moon in just the command-and-service module before the end of the year. They wouldn't merely loop around the Moon and return. Apollo 8 would fire its service-module engine to slow its momentum to enter lunar orbit. Then, after circling the Moon ten times, it would refire the engine to return. No such flight plan had been proposed previously, since the lunar module's own onboard computer and life-support systems had been designed as redundant backups should any problems arise in deep space. The new plan eliminated that safeguard.

Slayton needed to know if Borman and his crew would accept the assignment. No astronaut knew every inch of the newly designed command module better than Borman, and he had publicly voiced his confidence in its workmanship and safety. If he accepted the mission, Borman and his crew would become the first astronauts to ride the Saturn V, the first humans to escape the gravitational influence of the Earth, and the first to approach within seventy miles of the Moon's surface. Borman had already begun considering whether his current Apollo mission would be his last. He'd been working away from home for nearly the entire past year. The earth-orbital mission with the new lunar module that he, Anders, and Lovell were training for didn't excite him when far more interesting Apollo missions were on the horizon.

It didn't take long for Borman to give his response. He welcomed the challenge and told Slayton he was certain his crewmates would welcome it as well. When Frank told his wife about the proposed mission, Susan gave him her support, reluctant to stand in the way of her husband's career. But privately, in the aftermath of the fire, her faith in the space program had been shattered. She had watched Pat White struggling to adjust to a new life without her husband and could easily imagine herself raising their two boys alone. Frank reassured her that everything would be fine, but anxiety and doubt became her ever-present companions. Publicly she presented a brave face to the world, while privately she was getting through the day with a few more drinks than usual.

Low's lunar-mission proposal was an option NASA would undertake only if the first piloted Apollo mission—scheduled to launch on a smaller Saturn 1B in October—gave everyone confidence that the Apollo command-and-service module was ready to fly to the Moon two

With less than five months to prepare for the first flight to the Moon, the Apollo 8 crew of Frank Borman, Bill Anders, and James Lovell undertook what was then the riskiest mission of the American space program. Shortly before the launch, a prominent American physicist publicly called for NASA to postpone the mission, believing the space agency was pushing its luck.

months later. As to the Saturn V, von Braun had conveyed word that the unstable-combustion problem that had plagued the uncrewed Apollo 6 test flight had been eliminated. He now considered it moon-shot-ready.

The audacious lunar-orbit plan startled many in NASA's upper management, but their initial wariness soon subsided. Manned Spacecraft Center director Robert Gilruth argued that if NASA wasn't prepared to undertake this flight before the end of the current year, then they certainly shouldn't be considering attempting to land on the Moon less than twelve months later.

NASA's new deputy administrator, Thomas O. Paine, favored the daring idea, but Low's proposal had arisen while James Webb and George Mueller were attending a space conference in Vienna and out of contact with NASA's offices. Perhaps sensing that someone might attempt to instigate a controversial idea during his extended two-week absence, Webb had told Paine before his departure not to approve any major decisions until his return.

Webb was tired after running NASA for almost eight years, and the investigation following the fire had left him wary of what may be ahead. During a meeting the previous year, Lyndon Johnson had told Webb in confidence that he would not run for a second term as president. Webb resolved that he too would step down when Johnson left office in early 1969. Paine, a forty-six-year-old manager and engineer with General Electric, had been brought in as NASA's deputy administrator in early 1968 following Robert Seamans's resignation after the Apollo fire. Webb believed Paine understood the challenges facing NASA as much as he did, and Paine arrived knowing that he might be needed to provide administrative continuity during the coming Apollo lunar missions and under the new president to be elected in November.

But during the early discussions about the proposed Apollo 8 flight plan, Paine realized the timeline was so tight that the creation of computer software for the lunar approach and return had to begin immediately, before Webb and Mueller were even briefed about the December mission. Paine, a Navy veteran, chose to disobey Webb's request not to approve any big initiatives in his absence and authorized the development of the software, a decision he later likened to the choice Vice Admiral Horatio Nelson faced when serving as second-in-command of the British fleet at the Battle of Copenhagen. Nelson had held a spyglass to his blind eye and refused to acknowledge his superior officer's signal flags ordering retreat, a decision later celebrated by British naval historians.

Webb and Mueller were finally briefed about the Apollo 8 flight plan a few days later, over a secure phone line at the American embassy in Vienna, and Webb was alarmed. He thought Low's idea was bold but fraught with incredible risk. And he was certain if a second fatal space disaster occurred under his watch as NASA administrator, negative public opinion was likely to endanger all he had worked to achieve over the past seven years, including the lunar program. But rather than curtail all discussion, Webb asked that no word about it be uttered publicly until there had been a detailed assessment of Apollo 7's shakedown flight, scheduled to fly in a few weeks. Should the spacecraft appear suitable for a lunar voyage, then he would not oppose Low's flight plan.

Given Paine's daring action, Webb realized his deputy was willing to fully support the Apollo 8 lunar plan and suffer any repercussions. But Webb worried about possible scenarios and what might happen to the space agency if either Apollo 7 or Apollo 8 encountered serious trouble or ended in tragedy.

After his return from Europe and as plans for Apollo 8 moved forward in secret, Webb met with Lyndon Johnson at the White House to strategize continuity for the Apollo program. In the course of the meeting, Johnson recommended that Webb resign as administrator before the launch of the first crewed Apollo mission, reasoning that if a setback occurred on one of the early Apollo flights, Webb could carry on the fight in Capitol Hill's back rooms to ensure NASA's long-term future. It was an option Webb hadn't considered, even though he was planning to leave in 1969. Johnson would nominate Webb's selection of Paine as the acting administrator and he would be in place, overseeing the agency, when the new president arrived in January. If the risky Apollo 8 mission proved a success, the new president would want

Paine to continue running the agency through the first lunar-landing missions. They reasoned that the new president was likely to avoid appointing a new permanent administrator until 1970.

A few minutes later, President Johnson and Webb appeared outside the White House, where Johnson announced Webb's departure from NASA and Paine's nomination. In his comments to the press, Webb offered a word of caution. Though the first piloted Apollo mission would fly in a few weeks, he warned, the Soviets appeared to be "proceeding without letup" and could surpass the United States in space once again if the nation was not vigilant.

As Webb spoke on the White House lawn, a mysterious Russian spacecraft named Zond 5 was headed toward the Moon. Despite its ambiguous name, intelligence sources had concluded that Zond 5 was in fact an automated Soyuz spacecraft following the prepared flight plan for a future piloted circumlunar mission. On board were the first biological life forms to escape the influence of the Earth's gravity: two tortoises, an assortment of mealworms, wine flies, and some plants. Two days later Zond 5 whipped around the Moon in a free-return trajectory and headed back toward Earth. But when it began its reentry, a retrofire malfunction caused it to splash down in the Indian Ocean instead of landing in its intended recovery zone in Kazakhstan. After the spacecraft was successfully recovered, the animals were returned to the Soviet Union, where it was discovered the tortoises had lost about 10 percent of their body weight but showed no loss of appetite.

BIG SECRETS ARE difficult to keep quiet, especially when scores of people need to share them. A little more than a month after Borman had been offered the lunar mission, *The New York Times* cited reliable sources at the Manned Spacecraft Center discussing the possibility that "three Apollo astronauts may spend Christmas Day circling the Moon." No one would go on the record, especially as the Apollo spacecraft had yet to prove itself in flight.

Enthusiasm for the American space program was continuing to wane in a year of violence, protest, and discord, both in the United

States and around the world. Days before Apollo 7's launch, a prominent newsmagazine called Apollo a "program in decline . . . an embarrassing national self-indulgence." NASA's public-affairs office had hopes that a controversial four-and-a-half-pound addition to the Apollo spacecraft might help to rekindle public support: a black-and-white television camera.

Neither the spacecraft engineers nor the majority of the astronauts wanted the camera on board. The equipment would add additional weight to a spacecraft in which every ounce had to be justified. The astronauts feared it would dangerously distract from more important tasks. But Julian Scheer waged a concerted battle to get it approved by arguing that American taxpayers were entitled to see what they had been paying for.

The Apollo 7 crew, commanded by Wally Schirra, had been given a full ten-day schedule to put the new spacecraft through a series of shakedown tests in earth orbit. The television broadcasts were very low on Schirra's list of objectives, and his refusal to turn the camera on during the first day of the mission prompted a row with Houston. But when the crew finally consented to try out the camera early on the morning of the fourth day in orbit, it briefly diverted the world's attention from the war in Southeast Asia, the Soviet invasion of Czechoslovakia, the Summer Olympics in Mexico City, and the ongoing presidential campaign. The picture from the RCA camera was blurry and jerky but provided living room astronauts a vicarious opportunity to join in the mission in a way they never had before.

Shortly before Apollo 7's liftoff, the huge crawler-transporter holding a Saturn V with the Apollo 8 spacecraft dwarfed at its apex had departed the Vehicle Assembly Building for Pad 39A. If Apollo 7 encountered any major problems, Apollo 8's mission would be altered to remain close to home, albeit in a much higher orbit.

Within hours of Apollo 7's splashdown, NASA declared it "a near-perfect mission." Press speculation once again focused on Apollo 8's flight plan, which would be announced in a few days. But before NASA held its press conference, the Russians returned to space with their first piloted mission since the death of Komarov in Soyuz 1, a year and a half earlier. Coming shortly after Zond 5's circumlunar flight, Soyuz 3 was watched carefully in the hope that it might reveal Russia's long-term objectives. Carrying a single cosmonaut, Soyuz 3 met up with a second, unoccupied Soyuz vehicle. But when the two spacecraft attempted to link up, the docking was unsuccessful. Western space watchers were left uncertain whether this had been a training exercise for a future lunar mission or the eventual construction of a permanent space station.

Only a few days after Richard Nixon was declared the winner of the 1968 presidential election, NASA's now acting administrator, Tom Paine, finally made the official announcement: Apollo 8 would head to the Moon in late December on the most ambitious—and the riskiest—space mission ever attempted. He admitted it was entirely unclear whether the Soviet Union would undertake a similar mission as well, but it remained a distinct possibility.

Within hours of Paine's dramatic statement, Russia launched another Soyuz, Zond 6, on a circumlunar mission. But this time the biological specimens died when the cabin depressurized during the return voyage, a detail unreported to the world press at the time. The reentry was equally problematic, ending with a parachute failure and crash landing. Nevertheless, the Russians made every effort to save face by celebrating Zond 6 as a success. The Soviet Union then commenced a disinformation campaign designed to convince the world that Russia would attempt a piloted circumlunar mission in December. In fact, no such plans were being considered. In addition, a KGB colonel stationed in the United States was instructed to plant rumors at Cape Kennedy that Apollo 8's Saturn V had been sabotaged. It was hoped that this report would force a sudden launch cancelation. Fortunately, the KGB plan flopped when the letter detailing the sabotage plot was placed in NASA's crank-letter file. (It was only rediscovered when the KGB agent defected and confessed many years later.)

FRANK BORMAN HAD been given only four and a half months to get the Apollo 8 crew ready. Under normal circumstances they would have

trained for a year and a half. In addition to preparing for the lunar portion of the mission, as the first astronauts to ride the Saturn V, they
were put through a series of simulations intended to ready them for
every imaginable kind of launch emergency.

If for some reason the Saturn V should explode during launch, the
command module was equipped with an escape tower system powered
by a small solid rocket, which would separate and lift the spacecraft
away from any conflagration. But following the Apollo fire, additional
safeguards had been instituted. If there was a danger of combustion
prior to launch and the escape tower was not an option, the astronauts
had other alternatives. They could exit the spacecraft and access a
slide-wire harness mechanism that would rapidly bring them to a shelter, or they could jump down a chute and enter a bunker equipped
with padded chairs and shock absorbers intended to withstand an explosion. Both were unlikely options. Bill Anders dutifully carried out
the training, but privately he voiced his belief that in such a situation
there was little chance he would still be alive to make it to the doorway
of the blast room in time.

Anders did his own pragmatic and analytical risk assessment of the
mission. He figured there was a one-third chance they would not come
back; a one-third chance the mission would fail but they would return;
and a one-third chance it would go as planned. A decade earlier he had
routinely flown a nuclear-armed F-89J Scorpion from an American
base in Iceland, occasionally shadowing Russian bombers near the
Arctic Circle. Comparing the two assignments, Anders believed riding
the first Saturn V to the Moon was far less risky; to him, Apollo 8's odds
were acceptable.

What worried him was something else. Like nearly every other
fighter pilot he had met, Anders was most concerned that he would
make a stupid mistake. He would rather die than screw up.

Borman didn't harbor doubts about the upcoming assignment. He
simply told himself it would succeed. He had done all he could to reduce the mission to the essentials. To his mind it was a conservative
lunar flight plan; it didn't involve a rendezvous and docking or any extravehicular activity. When, at a Houston press conference, ABC's

Prior to becoming the first astronauts to fly the Saturn V, the Apollo 8 crew received emergency evacuation training in case of a launchpad explosion. Despite his assumption that there was little likelihood they would survive such a conflagration, in October 1968 astronaut Bill Anders familiarized himself with the newly installed slide-wire escape system at Pad 39B.

Jules Bergman pressed Borman for a comparative-risk equation, Borman had to admit spaceflight was a risky business but that only a minimum number of unknowns had been accepted for the Apollo 8 flight plan. For a risk comparison, he equated Apollo 8's mission to a combat tour in Vietnam.

It was a startling analogy at the time, but it captured how many astronauts with fighter-pilot experience viewed their unique situation. Anders felt embarrassed to be celebrated as a national hero in the press when largely unheralded American military pilots were flying from aircraft carriers, facing anti-aircraft artillery fire, and avoiding surface-to-air missiles on a daily basis.

The war in Vietnam was one of many issues dividing political opinion in the United States, but for those working long hours to fulfill Kennedy's deadline, there was little time to take notice. Borman had become so focused on helping NASA recover from the Apollo 1 fire and prepare for Apollo 8 that he had no opportunity to reflect on what was happening elsewhere in the world that year. The struggle for civil rights, the assassinations of Martin Luther King, Jr., and Robert Kennedy, the riots in the cities, and the political protests seemed as if they were happening on another planet.

LESS THAN TWO weeks before the launch date, the White House requested that Borman, Lovell, and Anders—then in the midst of intensive preparation—attend a black-tie dinner. This was intended as the departing president's formal tribute to all the astronauts and to James Webb, whom Johnson would honor at the dinner with the Presidential Medal of Freedom. The astronauts flew in from the Cape, while their wives arrived with a large contingent from Houston. More than 130 guests filled the dining area of the East Room, including Charles Lindbergh, his wife, writer Anne Morrow, Wernher von Braun, and the widows of two of the Apollo 1 astronauts. It was a lengthy evening, ending at midnight with a performance of excerpts from Jacques Offenbach's *A Trip to the Moon,* loosely based on Jules Verne's novel.

Anticipation about the coming lunar flight was mixed with a sense of melancholy, this being one of the last official dinners President Johnson would host before Richard Nixon assumed the presidency. The future course of the American space program would be set by another administration, and it was unclear what that might be. The only astronaut to speak that night was Wally Schirra, who toasted the departing president with a reminder that the United States could continue to set goals and go anywhere it wanted when those in charge provided vision and leadership. Though he didn't mention it that evening, Schirra hadn't forgotten the words Johnson had spoken to him in confidence a few months earlier. Johnson had been in Louisiana to deliver a morale-boosting speech to workers fearing layoffs at the Saturn V's first-stage assembly facility. The budget cuts to the space program were on everyone's mind. Turning to Schirra, Johnson said, "It's unfortunate, but the way the American people are, now that they have developed all of this capability, instead of taking advantage of it, they'll probably just piss it all away. . . ."

Competing with the formal speeches in the crowded East Room could be heard the noise of stifled coughs and sneezes. The Hong Kong flu had just been declared an epidemic by the National Communicable Disease Center, and upon their return to the Cape, the crew would enter semi-isolation, with limited physical contact with family. And so the dinner concluded with a surreal moment of last goodbyes. During his parting with his wife, Valerie, Bill Anders gave her an audiotape with a private message that was to be played should he fail to return.

Susan Borman continued to wear a brave face and hide her fears, even as Pat White and Betty Grissom sat a few feet away. On the day of the dinner, former Manhattan Project physicist Ralph Lapp was quoted in a newspaper story, calling for the delay of the Apollo 8 launch. "We are pushing our luck," he said. "NASA experts will assure you they have thought through the risks and have planned for them. Well, they didn't in Apollo 1." The *Tuscon Daily Citizen* published a headline: APOLLO 8 A DEATH TRAP? BORMAN DENIES IT.

In a year filled with much bad news, pessimists feared how it might end. The flight plan's many unknowns and dangers prompted journalists to speculate about what might happen if the service-propulsion-system engine failed to reignite in lunar orbit to bring them back to Earth. At a press conference, one reporter forthrightly asked Deke Slayton if the crew would have the option of taking suicide pills rather than wait until their oxygen was depleted. Slayton assured reporters that if such a situation arose, "The crew would have no recourse but prayer."

IN DECEMBER 1965, Susan Borman had taken her boys, Edwin and Frederick, to watch their father and Jim Lovell head into space for the first time. When Gemini 7's Titan II began to move upward into the Florida sky, she found watching the launch unbearable. As the boys' eyes followed the rocket's ascent, Susan, sitting between them, averted her eyes and buried her head on Edwin's shoulder while clenching Frederick's right hand. Three years later she decided not to go to the Cape. Of the three Apollo 8 wives, only Marilyn Lovell was in Florida.

Once again Susan would watch with Edwin and Frederick but this time in front of the living room television. Family and neighborhood friends came by to offer support and bring food, among them Pat White, who still lived a few doors away.

It was approaching 7:00 A.M. in Houston. Susan had situated herself on the carpet close to the television. Frederick and Edwin sat on the couch behind her. On the television, a countdown clock was superimposed over a telephoto image of the Saturn V. Jack King announced rather matter-of-factly, "We're coming up to the sixty-second mark on a flight to the Moon." Susan listened to King count down, but once again she couldn't look. She pulled her knees forward, pressed her fists into both cheeks, and tightly closed her eyes.

Inside the spacecraft, Borman, Lovell, and Anders could feel a vibration and sense a distant rumble as they became the first humans to lie atop a Saturn V as it started to release 7.5 million pounds of thrust. The experience was suddenly far louder and more violent than anything they had anticipated during the launch simulations. Bill Anders sensed the rocket moving so violently that he wondered to himself if perhaps everyone outside was witnessing a disaster. In his mind he pictured the small guidance fins located at the base of the first stage ripping into the girders on the launch platform. Looking at the control panel didn't offer him reassurance; the vibrations were so violent, his vision was little more than a blur.

Luckily, he knew that if they were in danger Frank Borman would twist the launch-abort handle. Located next to Borman's left hand, the handle would activate the launch escape-system tower. In fact, the violent buffeting had already prompted Borman to remove his hand from the abort handle to avoid grasping it to steady himself and while doing so unintentionally twist it in error.

Outwardly the launch looked good, with the Saturn V performing just as the Huntsville team had expected. Within twelve minutes, both the first and second stages had been discarded and Apollo 8 entered Earth's orbit, still attached to the Saturn's third stage, known as the S-IVB. In the course of their first revolution around the Earth, Anders was so occupied with monitoring the spacecraft's systems and making sure they were operating correctly, he had little opportunity to glance out the window. For Borman and Lovell, the view of Earth from space no longer held much novelty. Lovell was already the world's most traveled astronaut, having spent more than seventeen days in space. Borman briefly looked out the window and reported to Houston, "It looks just about the same way it did three years ago."

In Houston, astronaut Michael Collins was serving as the first of three rotating capcoms—capsule communicators—a term retained from the Mercury and Gemini eras. From his seat in Mission Control, Collins calmly informed the crew they should prepare to reignite their third stage while passing over the Indian Ocean during the second orbit. The five-minute firing would increase their speed to 35,505 feet per second and cause them to pull away from the Earth's gravitational influence, sending them on an elliptical trajectory toward the Moon. NASA's acronym-rich language allowed the moment to pass with little residual drama. "Apollo 8, you are go for TLI," Collins radioed, referring to trans-lunar injection. Borman's response was equally cool: "Roger, understand, we are go for TLI."

CBS's Walter Cronkite didn't want the significance to go unnoticed by his viewers. "That's the big decision! They are going to go for the Moon. There wasn't any excitement from Apollo 8. No cheers that we heard. That's the big one! . . . That undoubtedly was one of the most momentous—probably *the* most momentous—command ever given from Earth to a spacecraft in the seven years of the manned space program. And yet, as you heard, it was accepted with absolute calm." In the Mission Control room, Collins had little opportunity to reflect on his message's significance. It appeared to be just one step in a chain of commands. But Collins later admitted, "Jeez, there's got to be a better way of saying this!"

It was now time for Susan Borman to put on her public face and confront the battery of lenses and microphones on her front doorstep. The astronauts' wives morbidly referred to the gaggle of correspondents stationed outside their houses as "the Death Watch." Susan smiled, looked down, and answered the questions politely, once again accepting her assignment to support her husband and the space program. Privately, though, she couldn't shake an overwhelming premonition that Frank wasn't coming back. The newscasts had been filled with so much bad news that it seemed inevitable. She imagined how it would be told: Apollo 8 trapped in lunar orbit, circling the Moon for all eternity, carrying the bodies of three men who would never return. No one would ever look at the Moon the same way again. Late that evening when she was alone, she dutifully began to compose the words of a eulogy. She wanted it to be true to Frank's memory instead of words written by a government speechwriter who never knew him.

MISSION CONTROL, WHERE Michael Collins was acting as the first-shift Green Team's capcom, had served as the nerve center for every piloted spaceflight since the second Gemini mission in 1965. At the moment of liftoff, the Cape's launch team handed over responsibility for the mission to Houston. The average age of the Houston flight controllers during Apollo was roughly thirty. Some were engineering students who had come to work for NASA directly after graduating from college; others received their training while in the military. Nearly all were white and male, and their names were unknown to the American public, with the exception of Christopher Columbus Kraft, director of flight operations. Shortly after Mission Control began overseeing flights from the newly opened Manned Spacecraft Center, Kraft was featured on the cover of *Time* magazine. The disciplined organization and culture of the rotating Apollo Mission Control teams—each shift designated with a color, which became a mark of pride—evolved directly from Kraft's experience as NASA's first flight director.

The unique flight plan of Apollo 8 would require the addition of Mission Control specialists in entirely new disciplines that had not been part of previous earth-orbital missions. Teams had already been working for years designing and writing the computer programming to calculate the essential dynamics to safely achieve lunar orbit and the procedures to later return to Earth. One of the stars of the return-to-Earth team was a twenty-four-year-old mathematician who routinely put in seventy hours a week writing and refining the complex computer programs. Frances "Poppy" Northcutt was an employee of TRW, a leading defense and aerospace software and systems company contracted by NASA to handle the lunar calculations. A University of Texas math major, she was initially hired out of college as a technical aide assigned to analyze and graph data. (Her job title was "computress.") However, it soon became apparent she was overqualified, and she was promoted to the team calculating Apollo lunar return-to-Earth operations.

Northcutt had already taken courses in computer programming and celestial mechanics in college, so she was well prepared to work on designing the basic programming that would take the first humans to the Moon. She immersed herself in the challenge, seldom reluctant to ask questions in order to better understand every aspect of the problem.

Once Tom Paine approved the creation of software for the December lunar mission, Houston's team of mathematicians went into high gear. During the weeks that followed, Northcutt worked with Houston's flight controllers to familiarize them with the many intricacies of the return-to-Earth program, which was unlike anything they had worked with previously. Assigned to an operational support role at Mission Control, she was given her own station among the rows of young men wearing white shirts and headsets. As the launch date approached, Northcutt and her team were continuing to make tweaks and corrections to the computer program. The existing procedures dictated that no changes could be made to a program after a designated date, but in the case of Apollo 8 changes were being inserted well past the lock-in time and close to the launch date. Northcutt said it was not unlike a daredevil pilot flying an old biplane held together with baling wire and rubber bands, but if it worked, that was all that mattered.

Northcutt's very visible status as the first woman assigned a technical console in Mission Control also caught the attention of aerospace journalists looking for a story with an unusual angle. Though she was inevitably introduced to the world as "the first woman in Mission Control," the fact that she was young and photogenic also accounted for some of the media attention. But she took it all in stride, believing that if people could learn her story and see women holding important jobs in science and technology, it might inspire others and change the prevalent sexist stereotypes. Smiling for the media's lenses was a minor burden if it could better society.

THE FIRST DAY of Apollo 8's mission ended with an incident no one had planned for. As he tried to fall asleep, Borman became ill with violent nausea and diarrhea. Lovell and Anders were forced to undertake some unpleasant zero-gravity housekeeping by breaking out the few available paper towels and capturing the undulating blobs of bodily fluids floating throughout the cabin. If Borman had come down with the Hong Kong flu, all three were likely to be stricken.

Borman's condition went unmentioned over the publicly accessible audio communications link with Houston. Instead, the crew recorded a five-minute report about the situation, transmitted though a private, tape-delayed communications channel. A small panic ensued in Houston when the tape was accessed a few hours later and the NASA physician was informed of Borman's condition. Apollo 8 was already 125,000 miles from Earth, more than halfway to the Moon. But by the time Houston inquired about Borman's illness the next day, it had already subsided. Luckily, it wasn't the flu. Borman may have been experiencing what is now known as space-adaptation syndrome, a condition some astronauts undergo when adjusting to weightlessness in a roomier spacecraft.

The command-and-service module was positioned facing toward the Earth as Apollo 8 proceeded on its outbound journey. As a consequence, the astronauts never once saw the Moon during their approach. Out their windows they could only observe the Earth slowly receding in size. As early as the beginning of the spacecraft's third day in space, the Moon was exerting a greater gravitational influence on Apollo 8 than was its home planet.

In the course of preparing to enter lunar orbit, the spacecraft was positioned to fire the service propulsion engine. Curious, Bill Anders looked out his window and could see little more than a huge black void. The absence of stars made him aware that for the first time he was looking directly at the Moon, large, silent, and entirely unilluminated. This sudden realization caused the hair on the back of his neck to stand up. "Oh, my God!" he exclaimed.

Anders's words caused Borman a moment of concern; he'd assumed something on the instrument panel had caught Anders's attention. But as he looked up and saw his crewmates with their faces at the window, Borman gently said, "All right, all right, come on. You're going to look at that for a long time." Epiphanies could wait. Successfully carrying out their assigned mission was what mattered. Borman needed all three to focus on the engine burn that would place them in lunar orbit.

Just before Apollo 8 vanished around the far side of the Moon on the morning of December 24, the Black Team's capcom, Jerry Carr, conveyed a cryptic message: "Frank, the custard is in the oven at three-fifty. Over." Borman was puzzled for a few seconds before he realized this was a message of assurance from Susan. Years earlier they defined their partnership: He would be responsible for piloting the dangerous planes and spacecraft, while she would attend to the duties of a military wife, which she referred to as "staying home and cooking the custard."

Apollo 8 was scheduled to fire the large bell-shaped engine at the rear of the service module for close to five minutes shortly after disappearing behind the Moon. Should the engine fail to fire, Apollo 8 would reappear twenty-two and a half minutes later; if it fired successfully, slowing the spacecraft's momentum, it would reappear an additional ten minutes later, captured by the Moon's gravity in an orbit roughly seventy miles above the lunar surface. At Mission Control there was little the team of engineers could do but wait at their consoles, ready for any contingency.

On board the spacecraft was the most sophisticated miniature computer yet invented. The seventy-pound guidance computer, developed
at MIT, used inertial-navigation technology to define the spacecraft's
platform, monitor its position, and keep it pointed in the right direction. It was the first portable computer to employ an integrated circuit,
had a thirty-six-kilobyte memory, and operated on less power than a
sixty-watt light bulb. (Half a century later, a novelty greeting card's
microchip possessed greater computing power.) But if the timing of
the engine burn to place Apollo 8 in lunar orbit was too long or too
short, this state-of-the-art device was incapable of providing the calculations needed to make a correction.

Those essential computations took place in Houston, where a team
of engineers and technicians was standing by to analyze the data about
Apollo 8's position, velocity, and flight path. Using the cutting-edge
IBM System/360 mainframes in Mission Control's computer complex,
they would then produce updated or revised calculations when needed.
As a return-to-Earth program specialist, Poppy Northcutt was on hand
to analyze the lunar-orbital data as soon as the spacecraft's signal was
reacquired. If Apollo 8 emerged from behind the Moon earlier than
planned, its return-to-Earth trajectory would need to be checked, and
if necessary, refined. If the engine had burned too long, the spacecraft
could be in a dangerously low orbit and need calculations to refire the
engine to make an adjustment.

At the moment Apollo 8's signal was due to be reacquired, nothing
was heard. Neither telemetry nor voice communications were received. Had the engine failed to fire, Apollo 8 would have already reap-
peared on a trajectory heading back to Earth. In the Houston control
room, the anxiety was unbearable. Northcutt noticed no one was
breathing. Everyone just kept staring at the clock as seconds clicked
away. Capcom Jerry Carr was repeatedly putting out a call, "Apollo 8,
Houston. Over." Each time, only radio static came in return.

Inside the Apollo spacecraft, Carr's voice finally came through in a
crackling transmission received by the spacecraft's high-gain antenna.
"Okay, here we go," Lovell said to Borman and Anders when he heard
the call. At that moment Anders was attempting to photograph the
crater Tsiolkovsky, named for Russia's rocketry pioneer. "Go ahead,

Houston, this is Apollo 8," Lovell replied. "Burn complete. Our orbit is 169.1 by 60.5."

A cheer was heard in the Mission Control room. Northcutt and the team of specialists immediately began analyzing the orbital dynamics, determining the calculations to make it uniform. The spacecraft's lunar orbit had been unusually affected by "mascons," strange irregular concentrations of mass beneath the Moon's surface; the phenomenon had only been discovered earlier that year. In the case of Apollo 8, the mascons' influence on the spacecraft had proven so powerful that it altered the orbit sufficiently to extend the period of silence just prior to the reacquisition of the radio signal.

After relaying the data about the engine burn and other spacecraft systems, Carr asked the question on the mind of nearly every person listening. "What does the ole Moon look like from sixty miles?"

Lovell's response was immediate and precise. "The Moon is essentially gray," he said matter-of-factly. "No color. It looks like plaster of Paris or sort of a grayish beach sand." Those hoping for something a bit more romantic or surprising may have been disappointed, but his words still conveyed an undeniable sense of wonder.

Ever the business-minded commander, Borman focused on completing the ten lunar orbits and heading home. "While these other guys are all looking at the Moon, I want to make sure we've got a good SPS [service propulsion engine]," he called out to Houston. "And we want a 'go' for every revolution."

By the beginning of the fourth revolution, the Apollo spacecraft was still pointing down, facing the lunar surface in its established circular orbit sixty miles above the Moon. Anders, whose flight designation on Apollo 8 remained lunar module pilot even though there was no lunar module on this mission, had been given the additional assignment to oversee the extensive photography of the lunar surface. NASA scientists had narrowed down a list of potential landing sites and wanted the crew to obtain images of greater quality than what had been returned by the recent orbiting robotic mapping probes. In the weeks before the launch, Anders and Lovell had taken a crash course on lunar topography and geology and acquainted themselves with the prominent landmarks. After spending a few hours shooting photographs of the lunar surface, Anders began to find the Moon's landscape monotonous and a bit tiresome. Nearly all the craters began to look alike.

Fortuitously, Borman needed to readjust the spacecraft's attitude, because the ground stations on Earth were having difficulty acquiring a good signal. As he did so, he turned Apollo 8 so that it was no longer pointing downward. Anders, who was by his window, suddenly saw something he wasn't prepared for.

"Oh, my God! Look at that picture over there! Here's the Earth coming up. Wow, is that pretty!" Anders grabbed his camera with black-and-white film and shot the first picture of the Earth emerging on the lunar horizon.

By this point in the mission, Borman had begun making fun of Anders's habitual attention to doing everything in accordance with the flight plan. "Hey, don't take that," he teased. "It's not scheduled."

Borman's comment revealed one of the greatest ironies of the Apollo program. For all the rigorous planning for the first lunar mission, no one at NASA had bothered to consider the importance of the view of Earth they saw out their window. It had never been discussed in advance, nor had it been mentioned that the Earth would appear to them as if it was rising above the lunar surface in the darkness of space. The crew had been given no photographic instructions for taking a picture of the Earth, and they had no light meter to help prepare them to record the sight out the window.

Anders rapidly scrambled to find a color-film magazine. He was holding the one Hasselblad camera with the big 250-millimeter lens. That ungainly long lens had already been making Borman nervous whenever Anders moved it around in the cramped confines of the command module. "Just grab me a color exterior. Hurry up!" Anders called to Lovell, who was also searching for color film. Once the color magazine was attached to the Hasselblad, Anders had time to shoot only two images. By this point, Borman was also looking out the window. "Take several of them," he urged.

The unplanned photograph became known as "Earthrise," one of the most influential and iconic images of the twentieth century. Anthropologist Margaret Mead declared that the entire expense of the Apollo program was justified by the creation of that single picture. Visually, it captured the isolated fragility of the jewel-like Earth set against the barren and monochrome lunar surface.

Its widespread reproduction would occur as environmental consciousness and the philosophical concept of "spaceship Earth"—the realization that our planet was a contained, fragile, and finite system—were gaining currency in the larger culture. While many images of the Earth in space were taken subsequently during the Apollo program, no other photograph would have the same immediate and profound cultural impact as "Earthrise" when it was first seen throughout the world.

"Earthrise" was the first of two significant cultural moments that occurred in lunar orbit on December 24. The second took place on Apollo 8's ninth orbit.

In the midst of the crew's frenzied preparations a month earlier, Borman had received a telephone call from Julian Scheer. NASA's public-affairs chief said he understood the flight plan had scheduled a live television broadcast from lunar orbit on Christmas Eve. Borman had long opposed television broadcasts being added to Apollo missions, but after Apollo 7 there was no going back. The broadcast from the Moon would be seen by millions around the world and might result in the largest audience of all time, Scheer told him. He wouldn't give Borman any suggestions of what to say. Instead, he merely recommended that the crew say something appropriate to the occasion.

Drafting a speech was last thing Borman had time for as he readied for the journey. Neither Susan Borman, nor Bill Anders, nor Jim Lovell had had any good ideas. So Borman contacted an old friend from the U.S. Information Agency, who in turn passed the assignment on to his close friend Joe Laitin, a journalist who had recently stepped down as assistant White House press secretary. After handing it off, Borman gave it no more thought.

Laitin struggled to find the right words. He tried to imagine what it might be like in lunar orbit, on Christmas Eve, 239,000 miles from Earth. After two terrible assassinations, violence in the streets of Chicago during the Democratic convention, and countless conflicts all over the globe, "Peace on Earth" seemed a fitting message. But Laitin was certain it would be criticized as a hypocritical sentiment when the United States itself was waging an unpopular war in Southeast Asia. He tried to craft the perfect message but found the assignment the most difficult of his career. When his wife, Christine, found him well past midnight seated at the kitchen table before his typewriter, he was in a state of despair and out of ideas. She offered one suggestion, something both universal and fundamental that she thought might fit the occasion. Laitin thought it was an inspired idea, and he typed it up and sent it on the next day.

When Borman received the memo with Laitin's suggestions, he showed them to Lovell and Anders and asked that the notes be retyped onto fireproof paper and inserted in the flight plan.

Apollo 8 was in its ninth orbit, on the far side of the Moon, when Borman decided to focus on the upcoming broadcast. It was only a few minutes before the spacecraft would regain contact with the Earth. Recalling Scheer's words, Borman told Lovell and Anders, "We've got to do it up right, because there will be more people listening to this than ever listened to any other single person in history." With his ball-point pen, he marked the sections on the page that each of them would speak.

Looking at Lovell, he said, "Let Bill say the first four, and you say the next four, and I'll say the last two. There's no more." Borman wanted to maintain as much control of the broadcast as possible. Following the suggestion in the prepared notes, the camera would be pointed out the window and he would choose to end the broadcast at the moment when it felt right. But first they would provide their personal impressions from the past day and try to point out a few geographic features, including a possible future landing site on the Sea of Tranquility.

The television camera was attached to a window-bracket mount to provide a view of the lunar surface as it passed below, heading toward the approaching darkness on the distant horizon. As the terminator separating the sunlit portion of the Moon from the portion shrouded in darkness got nearer, the crew prepared to read. They had no reservations about the chosen passage. All were active in their local churches—Borman and Lovell were Episcopalian, while Anders was Roman

Catholic—but they didn't consider it a religious text per se. Anders thought the opening lines from the Bible's Book of Genesis an inspired choice, since the creation story is common to all the world's religions. And it would speak to the gravity of the occasion better than any words they might compose.

Holding the flight-plan binder with the marked page in his hands, Anders began the conclusion of the broadcast. "We are now approaching lunar sunrise. And for all the people back on Earth, the crew of Apollo 8 has a message that we would like to send to you:

In the beginning, God created the Heaven and the Earth.
And the Earth was without form and void,
and darkness was upon the face of the deep.
And the spirit of God moved upon the face of the waters,
and God said, "Let there be light."
And there was light.
And God saw the light, that it was good,
and divided the light from the darkness.

Anchoring the live broadcast for CBS News, Walter Cronkite was among the estimated one billion viewers watching in silence. As Anders began the reading, Cronkite momentarily thought to himself, "Oh, this is a little too much . . . this is corny." A second later Lovell began reading.

And God called the light Day,
and the darkness he called Night.
And the evening and the morning were the first day.
And God said, "Let there be a firmament in the midst of the waters.
And let it divide the waters from the waters."
And God made the firmament and divided the waters which were
under the firmament from the waters which were above the firma-
ment.
And it was so.
And God called the firmament Heaven.
And the evening and the morning were the second day.

Borman took the binder and began reading the bottom of the page, as the last sliver of light on the lunar surface was rapidly disappearing from view.

And God said, "Let the waters under the Heavens be gathered to-
gether into one place.
And let the dry land appear."
And it was so.
And God called the dry land Earth.
And the gathering together of the waters he called the seas.
And God saw that it was good.

"And from the crew of Apollo 8, we close with good night, good luck, a Merry Christmas, and God bless all of you—all of you on the good Earth."

A second later Borman gave a signal for Anders to cut the television transmission, and screens around the world suddenly broke to an image of video static.

On board the spacecraft, but unheard over the communications loop, Lovell said, "That's it."

Borman prompted Lovell and Anders not to speak another word over the communications link to Houston. Concerned that Anders might restore the television signal, Borman told him, "No. Leave it off. Great. Great!"

Despite only a minimum of preparation, Borman realized, the broadcast had ended with perfect timing. "Hey, how can you beat that? Jeez, we just went into the terminator right in time."

On Earth, everyone absorbed what they had just witnessed. Cutting away from the snowy screen, the CBS News director showed Walter Cronkite atypically wearing his glasses onscreen. Cronkite attempted to regain his composure as he looked into the camera. "Well . . . quite a finish for this last transmission from the Moon."

Apollo 8's Christmas Eve broadcast had preempted the network's prime-time evening schedule. In a bit of network housekeeping, Cronkite informed his viewers that in a few minutes they would resume their announced programming with a new episode of 60 *Minutes,* a CBS program that had premiered three months earlier. Tonight's holiday episode would feature Christmas Eve conversations with two women widowed earlier in the year, Coretta Scott King and Ethel Kennedy.

Throughout the United States, the Genesis broadcast was viewed in living rooms illuminated by Christmas trees and holiday decorations. The family of novelist William Styron was celebrating Christmas Eve with composer and conductor Leonard Bernstein and his family at their rural Connecticut farmhouse. This had become an annual tradition for the two families. Styron had been awarded the Pulitzer Prize for fiction earlier that year; Bernstein was at that moment unquestionably the most famous name in the American classical-music world. But like many intellectuals and artists of the late 1960s, Bernstein had no interest in the American space program, dismissing it as an expensive technocratic boondoggle. Styron was among the few who wanted to follow the progress of Apollo 8 on television that evening, something his host reluctantly agreed to allow.

As Anders began his reading, the holiday conversation and the laughter in the room grew quiet. Partygoers furtively looked toward the television to see what was going on. "I remember that a chill coursed down my back and an odd sigh went through the gathering like a tremor or a wind," Styron recalled. He looked over at his host "and saw on his face an emotion that was depthless and inexpressible."

In Mission Control, astronauts Neil Armstrong, Buzz Aldrin, and Michael Collins were seated next to the Maroon Team capcom, Ken Mattingly, as the broadcast proceeded. In the back of the room, Chris Kraft, George Low, Robert Gilruth, and Julian Scheer were watching as well. Shortly after the broadcast had concluded, Scheer took a call from an Asian journalist covering the flight in Houston. The reporter needed to file his story in a few minutes and was in a panic, having been unable to write down everything the astronauts had said.

"We heard the astronauts read something. Can you give me the words over the phone from the mission transcript?"

Scheer asked him, "Where are you calling from?"

"From my hotel room."

Scheer then calmly instructed him, "Open the drawer of the table next to your bed. In it you will find a book. Turn to the first page. The words you are looking for are there." Astounded, the journalist thanked Scheer for being so accommodating.

The American reaction to the Genesis reading was overwhelmingly favorable. However, some misinterpreted the intent of the reading as a provocative endorsement of American religious faith's superiority over communism. At the opposite extreme, atheist activist Madalyn Murray O'Hair commenced a very public court battle against NASA to prevent any religious observances on future spaceflights. O'Hair announced her lawsuit, arguing that the astronauts had violated the establishment clause of the First Amendment. In response, grassroots church groups flooded NASA with petitions of support. O'Hair attempted to take her fight to the U.S. Supreme Court, but it went no higher than the Court of Appeals for the Fifth Circuit, which ruled against her.

ON THEIR TENTH and final orbit, Borman, Lovell, and Anders readied their spacecraft for the decisive engine burn that would increase their velocity and push them out of lunar orbit. Now the calculations and preparations that Poppy Northcutt and others on the return-to-Earth team had formulated to bring Apollo 8 home would be put to the test.

The spacecraft was on the far side of the Moon, out of contact with Earth, when the crew reignited the large bell-shaped engine for slightly less than three and a half minutes and increased their velocity to 6,000 miles per hour. Once again there was no immediate sign of the spacecraft at the moment it was due to reappear. Capcom Ken Mattingly called out repeatedly, "Apollo 8, Houston," for two minutes, but the audio channel was silent. Fortunately, the eighty-five-foot dish antenna at Australia's Honeysuckle Creek ground station began to receive radio-telemetry data before any voices were heard. On the spacecraft, Bill Anders realized the high-gain antenna was in the wrong position and corrected the alignment.

"Please be informed there is a Santa Claus," Lovell was heard announcing to the world moments later. In Houston, it was only a few minutes past midnight on Christmas Day, and an immediate sense of relief filled Mission Control after a very tense twenty-four hours. An exhausted flight-operations director Chris Kraft immediately headed home. On the large display screen at the front of the Mission Control room, holiday colors surrounded a map of the Earth. Only now was it deemed the right moment to bring out a decorated Christmas tree, which was placed at the front of the control center.

It wasn't entirely evident to anyone at the time, but the space race between the United States and the Soviet Union came to a quiet close on that Christmas Day. Planning for a Russian piloted lunar program continued, but few believed it likely that a Slavic voice would be the first heard from the Moon, unless the Apollo program experienced another significant setback. With Apollo 8's splashdown on December 27, the Associated Press released another of its "space box scores," tallying up the United States's standing versus that of the Soviet Union. Some thought it trivialized an important moment in history, as if it were nothing more important than a sports championship.

The day after Apollo 8's splashdown, the editors of *Time* met to discuss their selection for the news magazine's annual "Man of the Year" cover feature. Weeks earlier, as 1968 was drawing to a close, the editorial board had tentatively chosen "the Dissenter" as the focus for their yearly feature. The past twelve months had been marked by continuous protests against war, militarism, repressive authoritarian governments, racial injustice, and academic censorship, and often answered by state violence and incarceration. *Time*'s editors acknowledged that dissent was motivated by an idealistic belief in a better tomorrow— a political vision of a future markedly different from what had been celebrated at the New York World's Fair four years earlier.

However, the successful conclusion of Apollo 8 prompted *Time*'s editors to reassess their plans and consider the significance of the just-completed space mission through a lens of history. Announcing that Borman, Lovell, and Anders would grace the cover as the magazine's "Men of the Year," *Time*'s editors acknowledged the important role of the dissenter as a force for change while suggesting that the flight of

Apollo 8 "shows how [the dissenter's] utopian tomorrow could come about." Apollo 8, the editors believed, would stand as a lasting example of what humans could do if, rather than turning inward or passively withdrawing from society, they chose to work together to challenge the unknown and face dangerous odds—even when reason argued otherwise.

The success of Apollo 8 also brought renewed attention to the hundreds of thousands of engineers, contractors, and scientists who were not represented on *Time*'s cover but had played a part in making history. NASA's new acting administrator awkwardly attempted to celebrate them by hailing the moon voyage as "the triumph of the squares." Despite her familiarity with a slide rule and the IBM mainframe, Poppy Northcutt was hardly anyone's idea of a square, and the success of Apollo 8 only heightened the media's attention to her role in the mission. But she was also increasingly aware of the responsibility that came with her fame. Were she to make a mistake, it might be used to justify the prevailing negative stereotypes about women in math and the sciences. She wasn't traveling aboard the Apollo spacecraft but she was another space-age pioneer, publicly breaking barriers and heading into uncertain territory.

Not long after the first features telling her story appeared on television and in newspapers, Northcutt started to receive fan mail. Girls and young women wrote her letters revealing that, before they saw her profiled, they hadn't been aware women were employed as engineers in the space program or worked at Mission Control. Her fame extended to the sorting rooms of the post office. One letter delivered to her home was simply addressed: "Poppy, Space Program, U.S.A."

Her employer, TRW Inc., decided to feature her in a national ad campaign, the first time the Fortune 500 corporation had chosen to highlight one of its employees in such an advertisement. She became the public face of TRW. Marketing one of the nation's leading defense and aerospace contractors as young, smart, and committed to changing the future was a bold departure from ads featuring images of missiles and submarines. And the campaign was a notable success. The positive response resulted in more fan mail—even proposals of marriage—

and an unexpected side effect: As TRW's most publicly visible employee, Poppy was almost impossible to fire.

Since she came to work on the space program, Northcutt had been irritated by a clause in the Texas labor law. Despite performing the same work as her male colleagues and working the same extended hours—often six days a week—she was paid 23 percent less. She found great meaning in working on Apollo, but the inequality of compensation was especially frustrating. Culturally, she observed that male engineers were always assumed to be competent unless proven otherwise. This was never the case with women, who were first required to demonstrate their competence.

Poppy Northcutt, who achieved fame as the first female flight controller in Mission Control during the flight of Apollo 8, appeared shortly thereafter in a successful advertising campaign for her employer, TRW, an aerospace, defense, and computer contractor.

Northcutt began to read about an emerging movement advocating for women's rights and for the elimination of sex discrimination in the workplace. "If anyone has the freedom to stick their neck out, it's me," she thought to herself. "And it's my obligation to do so, since I don't carry the same risk as the secretary who puts herself on the line." She learned about the National Organization for Women and their plan for a nationwide strike. When the date arrived, she told her boss she had decided to take the day off. She didn't tell him why, but the reason wasn't secret for long—she was recognized in a public demonstration outside a Houston federal building. *Life* magazine took notice as well and featured her in a cover story on an emerging phenomenon the magazine referred to as "Women's Lib."

Though *Time* magazine had chosen to alter the subject of its "Man of the Year" issue at the last moment, from within the story of Apollo 8 itself had emerged one of its "dissenters," motivated by idealism to effect change and promote equality though protest.

FRANK BORMAN HAD returned to Earth as one of its most famous citizens. He decided to stick with his decision never to command another space mission. Unlike Lovell and Anders, he had no great desire to walk on the Moon. Never again would Susan need to endure a week as stressful as the one she had just experienced. Now there were new demands, but of a safer sort. Within a month Borman and his family were sent on a tour of eight European countries, where they would be greeted by, among others, the pope, Queen Elizabeth, Spain's general Francisco Franco, and France's president Charles de Gaulle.

When Richard Nixon was sworn in as the thirty-seventh president of the United States three weeks after Apollo 8's splashdown, Frank and Susan Borman were given VIP seats only a few rows behind the podium. The nation was growing increasingly polarized, separated by racial and generational divisions exacerbated by political attitudes toward the war. But with Apollo 8's accomplishment, the country was reminded of another more promising possibility.

In his inaugural address, President Nixon quoted a line by poet and former librarian of Congress Archibald MacLeish, whose essay had appeared on the front page of *The New York Times* as the Apollo 8 astronauts were heading homeward. "To see the Earth as it truly is, small and blue and beautiful in that eternal silence where it floats, is to see ourselves as riders on the Earth together, brothers on that bright loveliness in the eternal cold—brothers who know now they are truly brothers."

CHAPTER SIX

MAGNIFICENT DESOLATION
(1969)

I**T WAS A** glorious New York tradition, which originated at the end of
the nineteenth century with the dedication of the Statue of Liberty.
Since then Manhattan had held more than 160 ticker-tape parades to
celebrate visits from kings, queens, generals, sports heroes, explorers,
and aviators. But by the late 1960s they were becoming a rarity. Nevertheless, when the Apollo 8 astronauts arrived in the city in early January 1969, there was no question: A ticker-tape parade was obligatory. It
was New York City's first in three and a half years.

More than two hundred tons of confetti and scrap paper were transformed into a chaotic celebratory blizzard by the force of the bitter
January winds rushing down the canyon of Sixth Avenue. Sitting for a
half hour in the numbing cold on an elevated limousine platform,
Frank Borman, Jim Lovell, and Bill Anders managed to smile and wave
at the hundreds of thousands of well-wishers standing on the streets
and peering from windows.

When the motorcade crossed onto Broadway, the astronauts noticed
that the avenue's signs had been changed to read APOLLO WAY. At the
United Nations, Secretary General U Thant welcomed the astronauts
as the world's first true universalists. "We saw the Earth the size of a
quarter," Borman responded. "There really is one world—we are all
brothers."

At precisely the same moment that New Yorkers were turning out in
the bitter cold to cheer and celebrate the epic achievement of Borman,
Lovell, and Anders, journalists were gathering in an auditorium in

Houston to meet another Apollo crew. Here there were no celebrations or marching bands. Instead, NASA was introducing the newly named crew of what was expected to be the first mission to attempt a landing on the Moon in just six months. Designated Apollo 11, it would undertake the landing as long as Apollo 9—the debut flight of the lunar module in Earth's orbit, in March—and Apollo 10—a test of the lunar module's systems a mere ten miles above the Moon's surface, two months later—were both successful. All three of the Apollo 11 astronauts were known to those who had covered the space program in recent years. The commander was Neil Armstrong, well remembered from the crisis on Gemini 8; the lunar-module pilot was Gemini 12's ace spacewalker Buzz Aldrin; and the command-module pilot was Michael Collins, a veteran of Gemini 10. Collins would remain in orbit above the Moon as the lunar module attempted its landing and return.

The festivities in New York overshadowed much of the news from the Houston press conference. And unlike the Apollo 8 crew, Armstrong, Aldrin, and Collins were fairly reserved when appearing in public.

No one could argue with Deke Slayton's selection of Armstrong to command this mission. He and Aldrin had just come from serving as part of the Apollo 8 backup crew, so their assignment to Apollo 11 followed Slayton's normal rotation. Michael Collins, who had been replaced on Apollo 8 by Jim Lovell due to his surgery, had since returned to flight status and been reassigned by Slayton to Apollo 11's third couch.

A perfectionist focused on obtaining every bit of valuable data, Armstrong personified the disciplined, conscientious test pilot. His calm, reticent, soft-spoken style was the antithesis of the macho jet-jockey stereotype. Still, Armstrong's handling of the emergency on Gemini 8 and his jet combat experience during the Korean War, while flying the experimental X-15 rocket plane, were legendary. But a crisis when at the controls of the experimental Lunar Landing Research Vehicle in May 1968 was arguably the most impressive demonstration of Armstrong's unique skills as a pilot. Nicknamed "the Flying Bedstead" due to its ungainly spindly appearance, the LLRV had been created to train

The ungainly and dangerous Lunar Landing Research Vehicle was nicknamed the "Flying Bed-stead." In May 1968, Neil Armstrong narrowly escaped an LLRV accident in Houston when his vehicle suddenly became dangerously unstable, causing him to fire his ejection seat when only 200 feet above the ground.

the astronauts who would land the lunar module on the Moon's surface. It was powered by a single large jet engine positioned vertically downward; sixteen smaller supplemental chemical thrusters allowed the LLRV's pilot to hover, maneuver, and land in a rough simulation of the Moon's one-sixth-gravity environment. But the LLRV was also a dangerously finicky contraption, and during what appeared to be a routine training flight at Houston's Ellington Field, Armstrong's vehicle began to gyrate wildly when only two hundred feet above the ground. As it pitched up at an angle, Armstrong miraculously fired his ejection seat and parachuted to safety, seconds before the LLRV crashed and burned not far from the Manned Spacecraft Center. Film of his accident and escape astounded all who viewed it.

Ambitious and brilliant, Aldrin was a fortuitous choice to serve as the lunar-module pilot for the first landing. His intense personality was nearly the opposite of Armstrong's, whose deceptive lack of ego could make him disappear in a crowd. During a meeting, Aldrin would make others aware of his extensive knowledge about an esoteric subject, often going into detail about things that were not the central focus of the discussion. He was the kind of person who enjoyed tackling theoretical challenges, tirelessly focusing his energies to arrive at the perfect solution.

During the Houston press conference, Aldrin revealed that his late mother's maiden name was Marion Moon. Kept from the press and the public, however, was a sad secret. His mother's death the previous year had been the result of an overdose of sleeping pills, and he blamed himself for her suicide. She had struggled with depression in the past, and as attention surrounding her famous son increased, so did her anxiety, which proved overwhelming. The social stigma surrounding mental health problems at the time was both powerful and pervasive, making any public mention of her suicide impossible. With *Life* magazine and NASA's public-affairs office highly invested in promoting the astronauts as perfect American role models, such sensitive personal details were kept confidential.

Despite their contrasting temperaments—Armstrong deceptively affectless and placid, Aldrin intense and focused—both were as competitive as the other alpha-male astronauts. A moment during the Houston press conference revealed some of the personal determination and drive that Armstrong kept largely invisible from the public.

From the gallery of journalists came the inevitable question that everyone wanted answered: "Which one of you gentlemen will be the first man to step on the lunar surface?"

Aldrin had performed his Gemini 12 spacewalk while commander Jim Lovell remained in the spacecraft. Such had been the protocol for all extravehicular-activity exercises during Gemini. It seemed logical to assume this practice would continue on Apollo. The first Apollo extravehicular activity to test the new self-contained lunar space suit would be performed on the next mission, by Apollo 9's lunar-module pilot. Aldrin had begun training for the lunar EVA and had reason to assume that, as the lunar-module pilot, he would be the first to set foot on the Moon, with Armstrong following him forty minutes later.

However, when Armstrong answered the question, he revealed, uncomfortably, "The current plan calls for one astronaut to be on the surface for approximately three-quarters of an hour prior to the second man's emergence. Now, which person is which has not been decided up to this point."

Following Armstrong's response, Aldrin diplomatically used one of the models of the lunar module to describe how each astronaut would exit the spacecraft when it was on the Moon. Aldrin, of course, also left open the question of who would be first man. Outwardly, neither man showed much emotion about the matter at the Houston press conference. But the tension in the room did not go unnoticed. Privately there was a standoff between Armstrong and Aldrin, as Armstrong made it clear he wasn't going to rule himself out of being first.

Fortuitiously, a solution arose out of practical necessity. The lunar module's designers at Grumman Aircraft had placed the hinge of the square exit door on its right side. When the door opened inward, the lunar-module pilot's movement within the spacecraft's cramped interior was confined; only the commander, on the left side of the cabin, had sufficient room to exit. Therefore, the commander would have to exit first and reenter last. The lunar-module pilot could not exit until he repositioned the door, moved to the left side of the cabin, and then exited through the hatch. Aldrin was justifiably disappointed but was professionally stoic whenever the issue was discussed with the press.

Although he received slightly less attention than his crewmates at the Houston press conference, Michael Collins emerged as the reporters' favorite. More socially at ease than his two crewmates, Collins possessed an observant eye, an understated wit, and a facility with language that distinguished him from many of the other astronauts. He memorably described the Apollo 11 crew as "amiable strangers," in contrast to others crews like Apollo 8, who bonded far more easily.

For Collins, the unique personal challenges were of an entirely different sort. By remaining in the command module in orbit above the Moon while his crewmates were on the surface, he would experience a form of solitude unknown to any human being in the entire history of the species. During forty-seven minutes of each lunar orbit, Collins would be cut off from all communication and any sign of life. On one side of the Moon would be the Earth, with its three billion inhabitants as well as Armstrong and Aldrin. On the other side would be Collins, entirely alone.

Aldrin had established himself as the astronaut specialist on orbital rendezvous and mission planning. With his test pilot's background,

Buzz Aldrin, Neil Armstrong, and Michael Collins, the "amiable strangers" of Apollo 11, photographed on the deck of a modified landing craft used for spacecraft water recovery training in the Gulf of Mexico. The preparations required the crew to practice changing into sealed garments, an attempt to keep the astronauts physically isolated in the event they returned to Earth infected with an extraterrestrial life form.

Armstrong was the astronaut office's expert on training vehicles and simulators. And Michael Collins's specialty was the Apollo space suit. The Apollo pressure suit was in fact a miniature self-contained spacecraft, with its own independent environmental system. It had to be bulky and insulated yet flexible enough so that the body inside could move freely and observe what was around it. It had to be strong enough to contain the interior air pressure in the vacuum of space, protect from micrometeorites, and withstand the harsh temperature differences on the lunar surface, which could range from 260 degrees Fahrenheit in the sunlight to -280 Fahrenheit in the shade. And it had to have a sustainable cooling system that would prevent the person wearing it from overheating.

Apollo 11's chosen landing site was a relatively level and smooth area in the Sea of Tranquility, near the lunar equator. It had been photographed from orbit by the Lunar Orbiter mapping probe and later by the Apollo 8 crew. Should the moonwalk proceed as planned, the Apollo 11 crew would deploy a small array of scientific experiments, including a seismometer, a device to measure the solar wind, and an optical reflector. It was hoped the latter would aid in determining the precise distance between the Earth and Moon within an accuracy of about three centimeters by using a laser beam sent from an earth observatory station.

In early spring NASA convened a Committee on Symbolic Articles Related to the First Lunar Landing. In his inaugural address, President Richard Nixon had said, "Let us go to the new worlds together—not as new worlds to be conquered, but as a new adventure to be shared." In light of those words, the NASA committee resolved that nothing during the lunar landing should suggest the United States was laying claim to the Moon's sovereignty. Such a claim would be a clear violation of the 1967 Outer Space Treaty, in any case, but the committee thought it necessary to avoid anything that could be misinterpreted as American imperialism.

The State Department endorsed the idea of displaying the flag of the United Nations, a proposal that was immediately rejected by the Nixon White House. As a compromise, the committee recommended that the American flag be planted in the lunar soil and that the achievement's larger human context would be conveyed by the unveiling of a commemorative plaque with a text that would be read aloud during the first moonwalk.

The committee would not make any recommendations about one aspect of the mission that was increasingly invested with suspense: what Armstrong should say when taking his first step on another world. The suggestion Scheer had given Borman to simply say something appropriate had proven to be good advice. With Armstrong, he did the same. Scheer briefly considered offering guidance but had come to believe that making any suggestions would be a mistake. He looked at the journals of Lewis and Clark and other explorers and concluded "that the truest emotion at the historic moment is what the explorer feels within himself, not for the astronauts to be coached before they leave or to carry a prepared text in their hip pocket."

Armstrong was admired for his emotional stability, lack of ego, and ability to make split-second decisions when under pressure. However, among his peers, he was one of the least compelling public speakers. His shy demeanor conveyed authenticity before cameras and microphones, but his words were seldom memorable. He appeared painfully uncomfortable on such occasions, and an awkward penchant for favoring engineering metaphors and jargon didn't help to forge an emotional connection with his listeners. He would now be placed in a situation that had already been imagined by scores of writers, filmmakers, and illustrators. *Esquire* magazine commissioned a cover story that asked, "What Words Should the First Man on the Moon Utter that Will Ring through the Ages?"

During the Apollo 11 mission, individual call signs would be needed for both the command-and-service module and lunar module whenever they were separated. For the first test flight of the ungainly lunar module on the Apollo 9 mission, its crew had given it the call sign "Spider." The more contoured command module answered to "Gumdrop." On Apollo 10 the call signs had been "Snoopy" and "Charlie Brown." However, for a mission as historically significant as Apollo 11, Julian Scheer thought the astronauts should adopt less frivolous call signs. Lyndon Johnson's former press secretary Bill Moyers, now the publisher of *Newsday,* publicly suggested naming one of the spacecraft in John Kennedy's honor. When his proposal reached the White House, President Nixon's domestic adviser, John Ehrlichman, summarily dismissed the idea as a partisan effort to perpetuate the name of Nixon's 1960 rival. Independently, Armstrong and Aldrin chose "Eagle" as the lunar module's call sign. Collins, however, was at a loss to come up with a name for the command-and-service module. During a phone conversation, it was Scheer who gently suggested to Collins, "Some of us up here have been kicking around 'Columbia' . . . "

JAMES WEBB HAD intended a smooth continuity of his NASA steward-ship on the eve of the first Apollo flight in late 1968. By endorsing Tom Paine as the agency's acting administrator, Webb and Johnson gave Richard Nixon little choice once he arrived in the Oval Office other than to officially nominate Paine to the permanent administrator's po-sition. With the first lunar landing scheduled in a few months, this was not the time to begin the search for new NASA leadership.

During the Kennedy and Johnson years, Webb had brought rare po-litical and corporate management experience to the NASA administra-tor's office. Now Paine would attempt to lead NASA through the Apollo program and beyond, but unlike his predecessor, he wasn't as experi-enced in the ways of getting things done in Washington. Though an active Democrat, Paine was primarily a scientist, engineer, and man-ager, and not used to spending long hours in conference rooms on Capitol Hill.

In contrast to Webb's congenial, confident, and assertive public presence, Paine was less imposing. His reedy voice, horn-rimmed glasses, and general affect brought to mind a stereotypical high school science teacher rather than the World War II Navy submarine officer and deep-sea diver that Paine had been a quarter century earlier. Nev-ertheless, in his new leadership position at the space agency he strove to rekindle the idealistic and aspirational emotions of the Kennedy era, reminding Americans of the promise of a better future that the Apollo program invoked. Upon the Apollo 8 astronauts' return from the daring mission he had personally approved, Paine echoed the rhetoric of early 1960s Camelot, telling the nation, "We should dream no small dreams. We should undertake those enterprises which lift the soul."

Incredibly, however, shortly before the scheduled launch of Apollo 11, Paine discovered that President Kennedy's surviving brother, Mas-sachusetts senator Edward Kennedy, did not view the Apollo program in the same frame of mind.

Paine had been at Cape Kennedy for the flight of Apollo 10 in May when he received word at his motel room that wire services were reporting that Senator Kennedy had called for a slowdown of the American space program. The comments had been made during a speech honoring rocketry pioneer Robert Goddard at Clark University. Sitting on the platform with Kennedy at the Goddard Library dedication were Goddard's widow, Esther, and Buzz Aldrin, whose father had been a student of Goddard's at Clark. Also in the audience was Wernher von Braun.

As they looked on, Kennedy proposed diverting funding from NASA's lunar program to address the many pressing problems on Earth: poverty, hunger, pollution, and housing. When he learned of the speech, Paine took Senator Kennedy's words as both a broadside against the space agency and a provocative snub at his brother's legacy. Delivering the words during an event honoring the father of American rocket science only deepened the psychological wound.

From his motel, Paine called Julian Scheer, and together they crafted a response. In the NASA rebuttal, Paine described Kennedy's words as a "disappointing and dispiriting vision of this nation's vigor and destiny in space." It also added an aside that an account of Kennedy's speech would not be transmitted as part of the daily news briefing read to the Apollo 10 astronauts on their way to the Moon.

In a nation already growing increasingly polarized about the war in Vietnam, Kennedy's words may have been intended for the ears of younger voters, many of whom viewed NASA and the space program as part of the greater military-industrial complex. Paine made it his mission to heal the rift with what he termed the "Kennedy establishment." He hoped he could dissuade the thirty-seven-year-old senator from any further comments about budgetary allocations for space, especially with the lunar landing approaching. To that end, Paine scheduled a lunch meeting with Kennedy in Washington.

The day they met, a leading Democratic politician was quoted in papers predicting Senator Kennedy would be the likely presidential nominee of his party in 1972. Maine Senator Edmund Muskie believed Kennedy was unstoppable, unless the political climate changed dramatically in the next three years. Should the responsibility for shaping the nation's space policy in the 1970s fall to another member of the Kennedy family, Paine hoped the new president would give more attention to sustaining the nation's spirit of adventure in space than acceding to a campaign platform designed to win votes with promises of NASA budget cuts.

Paine began the discussion by expressing his admiration for the vision and courage his brother had displayed when he chose to commit the nation to the lunar program. Both NASA and the nation were looking forward to fulfilling that decade-long challenge in a few weeks, he said. Therefore, Paine hoped the senator would refrain from any further critical comments during the Apollo 11 mission. He then asked if there might be some additional way to honor his brother during the moon mission. He suggested a symbolic gesture, such as leaving Kennedy's World War II PT-109 tie clasp or another personal item on the lunar surface.

In response, Senator Kennedy gave Paine no encouragement. In fact, the NASA administrator received the impression that the Kennedy family wanted little to do with Apollo 11. Kennedy revealed his belief that the Apollo program was more an aberration than a true reflection of his brother's lasting presidential legacy. And now that the Soviet threat in space had largely faded away, it wasn't a position that would motivate voters in three years. As to the upcoming launch, the senator said he didn't know if he would be there to attend.

THE CREW OF Apollo 11 made their last pre-flight public appearance during a thirty-minute live televised interview broadcast two days before their scheduled launch. Dressed in short-sleeve sport shirts and sitting in large easy chairs, Armstrong, Aldrin, and Collins answered questions over a closed-circuit video link so they could maintain an imposed semi-quarantine prior to the flight. The interviewers were situated in a building fifteen miles away.

The crew gave every indication that they had great confidence in carrying out their mission. Aldrin spoke of "when" the lunar module

While confined to a pre-launch quarantine, the Apollo 11 crew answer questions from jour-nalists during a live television news conference less than forty-eight hours before their depar-ture. Deke Slayton, the chief of the astronaut office, sits to the side, at left.

Eagle would touch down, not "if." Since the triumphant mission of Apollo 8, the two test flights of the lunar module—the first in Earth's orbit on Apollo 9, and the second descending to ten miles above the lunar surface on Apollo 10—had demonstrated that the equipment performed as expected. Armstrong appeared slightly nervous in the interview, choosing his words with caution as usual. However, he agreed with some space-agency estimates that there was an 80 percent chance that the mission would be completely successful and placed the crew's chance of returning safely even greater than that.

Among the veteran newsmen who had chronicled the space program for years, there were many who thought it unlikely the first attempt at a landing would fulfill Kennedy's goal. They had covered the Apollo fire, Gemini 8's emergency return, and any number of aborted launches. Apollo appeared to be on an extended lucky streak in the two years since the fire, but a harrowing moment during the Apollo 10 mission—when the lunar-module ascent stage gyrated briefly before being brought under manual control—reminded everyone how the un-expected was always a part of every flight.

Aldrin mirrored Armstrong's confidence and admitted the only thing about the lunar module he would have preferred was a slightly larger descent-engine fuel tank, calling it "a little bit of extra protection in the hip pocket."

Once again, Michael Collins was much more at ease in front of the cameras, reminding everyone that he was probably going to be the only American who wouldn't be able to watch his crewmates on live televi-sion as they walked on the lunar surface, "so I'd like you to save the tapes for me, please."

During their final news conference, Armstrong was also asked his thoughts about a Soviet scientific lunar probe Luna 15, which *Tass* re-ported had been launched the previous day. The purpose of its mission had not been disclosed, but it was believed to be an attempt to upstage Apollo 11 by soft-landing a robot on the Moon, which would obtain a sample of the lunar soil and then return it to the Earth. Armstrong said he thought the chance that Luna 15 would affect the planning of their mission was infinitesimal and he wished the Russians success.

The previous week Frank Borman had returned from a trip to the Soviet Union; he was the first American astronaut to make such a dip-lomatic visit. While in Moscow he had heard Russian space scientists alluding to an upcoming robot mission but was unaware that the final preparations were already under way. In recent years, a vocal segment of the American scientific community had advocated for less-costly, unpiloted research missions rather than human piloted flight. Should Luna 15 succeed, it would likely strengthen that position. Appearing on *Meet the Press* to discuss the upcoming Apollo 11 mission and Luna 15, Borman acknowledged that automated probes played a worthwhile role in space exploration, but there was still no substitute for human judgment during the course of a space mission. Since commanding Apollo 8, Borman had become one of the country's most admired citi-zens, honored for his courage, honesty, and straight talk. NASA and the Nixon White House asked him to serve as a special adviser to the president during the Apollo 11 mission, including soliciting his thoughts on the ceremonial aspects of the first moon landing.

Luna 15's lunar mission was of concern to flight director Chris Kraft. Unlike Armstrong, he couldn't dismiss the possibility that the two craft might affect each other's orbits. Just to be safe, Kraft called Borman at the White House and asked if he could use his new connections in the USSR to obtain details about Luna 15's trajectory. From his small White House office, Borman sent a teletype message to the head of the Soviet Academy of Sciences, using the now-famous hotline set up by President Kennedy in the aftermath of the Cuban Missile Crisis. In his message, Borman requested Luna 15's orbital parameters, in accordance with the rules of the 1967 Outer Space Treaty. Within hours Borman received a response, revealing that Luna 15's trajectory would not intersect Apollo 11's orbit, as well as an assurance that should anything change he would be notified promptly. This was the first cooperative message exchanged between the Soviets and the Americans during an ongoing space mission.

A spirit of camaraderie and goodwill had surrounded Borman during his recent Soviet visit, which came as a result of an invitation from Russia's ambassador. Both President Nixon and his National Security Council head, Henry Kissinger, had encouraged it. Borman's itinerary took him to ceremonial events and included a visit to Star City, the cosmonauts' village outside Moscow. He and Susan Borman placed wreaths on the tombs of Lenin and space heroes Korolev, Gagarin, and Komorov. But in spite of the red-carpet welcome and all the smiles, some things remained off-limits to a United States Air Force colonel. When Borman inquired whether he could visit the Baikonur Cosmodrome—the Soviet Union's launch facility, located in a remote region of Kazakhstan—his genial hosts awkwardly informed him that a visit, unfortunately, wouldn't be possible.

Not long after Borman's return, Western publications printed reports that offered a likely explanation for his hosts' discomfort. Borman's visit had coincided with a dramatic turning point in Soviet space history. Shortly after Borman's arrival in Moscow, a team of engineers 1,600 miles away attempted a night launch of the massive N-1, the moon rocket Korolev's team was working on when he'd died three years earlier. Unknown to Western observers, this was Russia's second test of the N-1. Five months earlier, another test had failed a minute after liftoff. When the Baikonur team made its second attempt, the N-1 encountered trouble seconds after ignition, as one of its thirty first-stage engines blew apart, setting off an irreversible and disastrous chain of consequences. The Saturn V's Soviet rival climbed into the sky for no more than twenty seconds before it slowly began to fall back to the Earth, tipping sideways in a tower of flame. It crashed to the ground, destroying the launchpad with an explosion equivalent to two hundred fifty tons of TNT—the largest man-made non-nuclear explosion in history. The secrecy enveloping the launch and its aftermath was so effective that Borman had no indication that the Soviet space program had endured so decisive a setback while he was in the country.

With Luna 15, the Soviets were making one last play for world attention while the international press was converging on Cape Kennedy. The mysterious little robot probe heading toward the Moon was providing the final suspenseful chapter in the saga of the lunar space race. It was a far cry from the heady days earlier in the decade when beaming portraits of Yuri Gagarin, Valentina Tereshkova, and Alexei Leonov appeared below huge headlines trumpeting the Soviet's latest first in space.

THE LAUNCH OF Apollo 11 would be a media event like no other in the history of the world. For months the suspense had been building. After nearly a decade of planning and anticipation, President Kennedy's goal appeared as if it might be achieved on time and close to budget. Television producers, magazine editors, and advertisers all knew they would have a unique opportunity to impress their customers. Editors commissioned special magazine and newspaper supplements designed to educate their readers with retrospective histories and details about the equipment and the crucial stages of the mission. In the media capitals of New York and London, television networks prepared features and interviews with poets and intellectuals to supplement their planned live coverage of the mission; Duke Ellington and Pink Floyd composed and recorded original musical performances that would be premiered on air.

At Cape Kennedy, NASA had issued nearly three thousand passes to credentialed journalists, more than eight hundred of them representing foreign media outlets, including many from Soviet countries. Upon arriving in Florida in the days before the launch, every reporter was hungry to discover a story that was distinctive and original. By now every motel owner in the costal region near the Cape had been interviewed about the scarcity of available rooms or had offered reflections on how their business had changed since Alan Shepard's first flight eight years ago. The presence of the countless cameras, microphones, and reporters assembled in Florida did not go unnoticed by veteran activists in one of the other great national historical turning points of the 1960s: the crusade for civil rights.

Before his murder the previous year, the decade's most influential advocate for nonviolent civil disobedience would occasionally reference the achievement of the American space program as an example of what could be accomplished by human resolve and determination. In his Nobel Peace Prize acceptance speech, Dr. Martin Luther King, Jr., contrasted modern technological marvels, such as thinking machines and spaceships, with the moral and spiritual deficiency that led to injustice, poverty, and war. During the last year of his life, when opposing the expense and immorality of America's "unjust, evil, and futile" military intervention in Vietnam, King would also question the nation's decision to spend billions to put a man on the Moon while failing to allocate billions toward putting "God's children on their two feet right here on Earth."

King's rare words of criticism did not question the space program's aspirational goals or its morality. Rather, he addressed it as a case of misdirected national priorities in a time of crisis. After King's murder, his successor at the Southern Christian Leadership Conference (SCLC), Ralph Abernathy, continued to work on King's last major effort: the Poor People's Campaign for economic justice. During the troubled summer of 1968, the Poor People's Campaign had driven a symbolic mule train of carts from Mississippi to Washington, D.C., and mounted demonstrations at both the Democratic and Republican conventions.

Less than twenty-four hours before the scheduled launch of Apollo 11, Tom Paine was attending a meeting with Kurt Debus, the director of the Kennedy Space Center. Shortly after noontime he was interrupted by the arrival of the center's chief security officer, who reported that access to the space center's main entrance was under threat from a planned demonstration. All the available deputy sheriffs in nearby counties were being marshaled to counter the approaching protesters, in an effort to keep the gate open. Paine and Julian Scheer immediately realized that if the local police engaged in a violent confrontation with demonstrators before the eyes of the world press, NASA would be facing a public-relations disaster that promised to tarnish its greatest triumph.

The demonstrators heading for Cape Kennedy were from the Poor People's Campaign, led by the Reverend Ralph Abernathy. Once again, to symbolize their cause and attract the attention of the media, a small mule train of wagons accompanied them. The SCLC had also just released a position paper describing the march as a demonstration intended to caution against the possible illusion that conquering space would solve the country's domestic problems. It opened by stating clearly that the SCLC's Poor People's Campaign did not oppose the launch of Apollo 11: "We hail this historic achievement." But it also asserted another lofty goal that merited attention: "America can and must end poverty, racism, and war on Earth."

Paine told the Cape's security chief that the sheriff's deputies must avoid a confrontation. He and Scheer then decided to handle the situation by themselves. During his career as a journalist, Scheer had covered the formative years of the civil rights movement, had befriended the writer James Baldwin, and interviewed many of the leading figures. He sent word that he and Paine would like to meet with Abernathy and members of the campaign for a public conversation that afternoon. Only the two of them would be representing NASA, and the press was invited to attend as well. In his reply, Abernathy suggested that they meet in an open dirt field near the gate. Paine and Scheer would stand on the north end of the field, Abernathy and the marchers would gather on the south end, after which they would both approach to meet at the center.

The humidity became increasingly oppressive in the early afternoon. At the south end of the field, the demonstrators gathered, some singing and clapping, others holding hand-lettered signs of protest. The chants were led by the booming voice of forty-three-year-old Hosea Williams, an iconic figure of the civil rights era whose biography included an escape from a lynch mob at age thirteen and facing down club-wielding state troopers on Selma's "Bloody Sunday."

At the north end of the field, Paine and Scheer wore traditional NASA attire: white oxford shirts, skinny neckties, and plastic-enclosed security identification badges. Looking to the horizon, Paine noticed thunderheads forming in the tropical air and heard the slight rumble of a gathering storm. A few hundred feet away, in the shadow of a Mercury Redstone rocket standing on display, he saw approximately one hundred marchers approaching, with Abernathy and Williams at the front. The press was assembling as well, thankful to have some action to report as they waited for the launch. Few could forget that little more than a year earlier, Abernathy had cradled the head of his closest friend, Martin Luther King, Jr., as he lay dying on the balcony of the Lorraine Motel in Memphis. He and King had drawn strength from each other for more than a decade, during which they had been arrested and jailed as collaborators seventeen times. Today he marched in King's stead, dressed in a buttoned-up suit, seemingly unaffected by the sweltering weather. Behind him walked women and children holding signs. The elderly followed next, with the mules, the carts, and a few dogs at the rear. To Abernathy's right, Paine noticed one marcher supported on crutches, struggling to keep pace. Near the halfway point in the field, the two NASA men could begin to make out the familiar poignant sounds of "We Shall Overcome," a song Scheer had heard sung countless times while covering tense situations for *The Charlotte News*.

News cameras were trained on both the column of marchers and the two isolated men who slowly approached from the opposite direction. Paine's earlier work on a federal American cities task force led him to believe that with the creative application of science and technology, many of the existing urban problems could be solved, though doing so would not be easy. Scheer, on the other hand, had begun to feel a balance of conflicting emotions during his time with NASA. He was proud of what the nation had accomplished in space during the past decade, during a time when so much had gone wrong elsewhere. However, his pride was mixed with a sense of guilt. He was sympathetic to the unavoidable question of national priorities that had brought the protesters and the world press to the field outside the space center.

Cameras and microphone booms circled as Abernathy and Williams reached out to shake the hands of Paine and Scheer. Two men representing the world of science, technology, and corporate and government power looked into the eyes of two African American men of deep religious faith and committed social activism. All four were close in age and had served their country during World War II. Yet they represented segregated cultures and philosophies. Their lives had been shaped by

NASA administrator Tom Paine smiles as public-affairs chief Julian Scheer (left) shakes the hand of the Rev. Ralph Abernathy during the Southern Christian Leadership Conference protest near the gate to the Kennedy Space Center on the afternoon before the launch of Apollo 11. Future District of Columbia congressman Walter Fauntroy holds a microphone near Abernathy while Hosea Williams (extreme right) looks on.

the country's troubled history and a legacy that continued to define American life.

Paine said he had come because he wanted to meet them, acknowledging he had read the SCLC's position paper calling for a national program to aid the nation's forty million poor. He expressed his hope that the space program would serve as a demonstration of what the country could do when it mustered its skills, resolve, and resources. It was his hope that it would also spur the United States to face up to its responsibility to do better at tackling the problems on Earth.

Abernathy reiterated that they were not at Cape Kennedy to protest or disrupt the scheduled launching. "We are happy because our nation is a leader in this area," he said, but then added, "We have mixed our priorities in our country." He spoke with pride of the nation's scientific and technological advancements, yet he said he could not express any pride that the nation failed to feed its hungry, clothe its naked, provide housing for the needy, or bring every citizen into the mainstream of American life. "For [a] nation to spend billions of dollars to put a man on the Moon and . . . not spend sixty dollars to stand one on its feet right down here on Earth. That is a sick nation."

He then made three requests: VIP passes to the launch for ten poor families; Paine's support for a crash program to combat hunger and poverty; and the assistance of NASA's engineers and technicians to work toward solving the nation's social problems.

Paine swiftly and happily granted the passes. "I only wish it were possible to do the other things as easily," he admitted. "To change men's hearts, to lift people from poverty—these are gigantic tasks beside which our moon program is child's play."

And then to make his point he stated, "If it were possible tomorrow morning to not push the button and solve the problems of which you are concerned, believe me, we would not push the button."

As the first raindrops began to fall, Paine and Scheer departed, walking toward the NASA car parked on the north side of the field. They could hear another round of "We Shall Overcome" growing fainter in the distance. Scheer turned toward his boss and exhaled a sigh of relief.

The press cameras remained with the demonstrators and kept filming while Abernathy closed his eyes and offered up a prayer for the
safety of the astronauts, with the hope that while heading to the Moon
they might also reflect on the nation's ten million poor and hungry.
Within a few minutes, film of the meeting was being processed for
inclusion on the evening newscasts. Back at the public-affairs office,
Scheer arranged for buses, guest passes, and meals for the one hundred members of the Poor People's Campaign, who would be at the
launch the next morning.

For a few minutes, an unkempt field outside the Kennedy Space
Center gate served as common ground where two opposing philosophies motivated toward achieving a better America met to listen to
each other.

ON THE MORNING of Wednesday, July 16, 1969, the three commercial
television networks were on the air early with live television coverage
from Cape Kennedy. All had individual broadcast booths with desks
that allowed their correspondents to appear on camera with a view of
Pad 39A and the Saturn V three miles in the distance.

Before sunrise, a few journalists with press passes assembled near
the exit door of the Kennedy Space Center's Operations Building to
catch a glimpse of Armstrong, Aldrin, and Collins as they boarded the
white transport van that would drive them to the launchpad. When the
trio exited the building, carrying their suitcase-size portable air-
conditioning and ventilation units, they waved to the small crowd a
few feet away. One eyewitness, aware that he was literally watching
three men in space suits begin their journey to the Moon, compared it
to "watching Columbus sail out of port."

For CBS News anchor Walter Cronkite, the coming week would
culminate a decade of preparation. He had been absent from his
evening-news anchor desk for nearly a month, laying the groundwork
for the historic broadcasts by studying NASA reports, spacecraft manuals, and flight plans as if cramming for a final exam. Unlike Chet
Huntley, his craggy counterpart on NBC, who openly questioned the

lunar mission and the nation's priorities that morning, Cronkite's enthusiasm and support for the space program was never in doubt. Culturally and politically, Cronkite's influence was unique in broadcast journalism. Within three years he would outpoll presidents and senators as "the most trusted man in America."

An hour before launch, network cameramen began to focus on recognizable faces arriving at the space center's press site and VIP viewing area near the Vehicle Assembly Building. Vice President Spiro Agnew arrived with a contingent from the White House. Since on orders from the Apollo flight surgeon President Nixon had been prevented from dining with the astronauts the evening before the launch, he remained at the White House, where he prepared to watch the TV coverage with Frank Borman, and made plans to greet the crew on their return.

The television cameras followed the arrival of former president Lyndon Johnson and Lady Bird Johnson. A few feet behind them was the recognizable face of General William Westmoreland, forever linked to Johnson's White House legacy and his controversial command of U.S. forces in Vietnam. Washington D.C.'s mayor, Walter Washington, scanned the bleachers filled with other politicians and celebrities and remarked to his wife that he saw very few black faces among the invited guests. In another part of the vast VIP area, however, stood Ralph Abernathy and the families from the Poor People's Campaign.

The networks had little to report about the Soviet Luna 15 probe heading toward the Moon, but the continued Soviet silence about its purpose added even more suspense to the unfolding news events. A leading British astronomer monitoring the probe had just reported that its trajectory indicated it would likely attempt to retrieve a lunar sample. Bemoaning the lack of international cooperation between the superpowers, he boldly predicted that within ten years—and after lunar bases had been established—the United States and the Soviet Union would realize it would be in their best interest financially and diplomatically to collaborate in space.

Meanwhile, the immediate suspense was heightening as the steady voice of the Cape's veteran public-affairs announcer Jack King chronicled each moment of the countdown. King was seated in a swivel chair before a monitor in the huge Launch Control Center. Like the hundreds of other technicians in the room, King wore dark slacks, a white shirt, and a headset. "The target for the Apollo 11 astronauts, the Moon, at liftoff will be a distance of 218,096 miles away," he reported. Few knew his name or his face, but for the past few years his familiar voice had accompanied nearly every launch. He liked to work without a script, relying on a few note cards and his extensive knowledge of each step that preceded the launch.

Beneath the three-story glass window that dominated the eastern wall of the control center stood Wernher von Braun, whose Saturn V would be put to the test a sixth time. He would occasionally train his oversize binoculars on Pad 39A, but he never betrayed any sense of anxiety.

Also carefully observing the activity was documentary filmmaker Theo Kamecke, the only person not directly involved in the launch who'd been given a pass to the control center that day. Commissioned by Julian Scheer to create a film record of this historic moment, Kamecke sensed the unease in the control center as the countdown clock clicked closer to zero. "I suddenly understood what it meant to smell fear," he recalled. "Every single one of those hundreds of people in the room was afraid that their gauge, or their valve, would go wrong—and the rocket would blow up."

In newsrooms across the country, a singular occurrence took place. Suddenly, less than two minutes before the launch, all were transformed by a pervasive silence. The automated UPI and AP wire-service machines, which normally provided an unending background clatter of banging type and ringing bells, were not making a sound. News dispatches from around the globe came to a halt while the world watched the live transmission from Cape Kennedy and listened to Jack King continue the countdown.

"Twelve, eleven, ten, nine, ignition sequence start. Six, five, four, three, two, one, zero." Though his voice remained calm, the emotion of the moment affected King's delivery, and he could be heard slightly stumbling his words as he said, "All engine run-ning." Quickly recovering his composure, he continued, "Liftoff! We have a liftoff, thirty-two minutes past the hour. Liftoff on Apollo 11."

A wave of bodies rose to their feet on the VIP bleachers, as everyone strained to get a better view and position their snapshot and 8mm movie cameras. Lady Bird Johnson beamed under a plastic cowboy hat. To her left, her husband watched the slow progress of the Saturn and thought that, in its struggle to escape Earth's gravity, it appeared as if it were being lifted by the consolidated physical strength of the half million Americans who had built it.

A few seconds after the billowing clouds at the base of the pad and the brilliant flame of the Saturn's engines became visible, the observers watching three and a half miles away heard a rolling and overpowering roar. With it came a buffeting concussive shock wave of pressure that could be felt through the ground and in the air pushing against their chests and eardrums.

The violent controlled combustion needed to lift 6.5 million pounds away from the force of the Earth's gravity was an unequaled engineering achievement—and remains unequaled half a century later. The Saturn V was the physical creation of the mind and hand of the human species, a work of imagination conceived by beings composed of living cells, beings who evolved from aquatic life on a small planet orbiting a minor star. The launch of Apollo 11 followed in the tradition of the early explorers who expanded out of Africa's Great Rift Valley and extended their presence to nearly every part in the world. Like Stonehenge, the Pyramids, the great cathedrals, and the statues on Easter Island, Apollo was an enterprise guided by the human genetic code, which over the course of hundreds of thousands of years had favored those who could dream, reason, persist, and create. A technological marvel, a manifestation of political will, or, perhaps, a work of cosmic conceptual art, Apollo 11 defined what it was to be human.

After the Saturn V had cleared the tower, King and everyone else in the Launch Control Center—whose job permitted it—swiveled in their chairs to watch the rocket climb into the Florida sky. Some rose to their feet and reached for their binoculars, following the Saturn as it slowly disappeared from view.

In his CBS News booth, Walter Cronkite tried to capture the immediacy of the moment for his viewers, describing the buffeting that had so surprised him at the Apollo 4 launch two years earlier. Now he

Former president Lyndon Johnson, flanked by Lady Bird Johnson and Vice President Spiro Agnew, observes the launch of Apollo 11 from the VIP observation stand. In the row behind them is James Webb, NASA's administrator during the Kennedy and Johnson years. During his entire tenure running NASA, Webb attended only one piloted launch from Cape Kennedy. He explained, "I had a job to do in Washington."

mentioned the physical shaking in calmer tones, but he could still hardly contain his excitement. "What a moment! Man on the way to the Moon!"

Once Apollo 11 had disappeared from visual range and achieved orbit, journalists in the VIP viewing area scrambled to obtain the thoughts and reactions of some of the famous names before they departed.

"I really forgot the fact that we had so many hungry people," Ralph Abernathy told a UPI reporter. "I was one of the proudest Americans as I stood on this spot," he said, adding, "This is really holy ground."

Nearby, another journalist cornered outspoken segregationist Georgia governor Lester Maddox as he stood surrounded by a contingent of state troopers. Maddox had risen to fame as an Atlanta restaurateur who refused to serve African American patrons. "Phooey on Ralph Abernathy!" Maddox shouted, certain his often-used catchphrase would appear in the next day's papers. "Let's quit feeding the have-nots. Apollo 11 would never have happened if it wasn't for free enterprise."

Few in the media overtly noted it, but prominent in his absence was Senator Edward Kennedy, who remained in Washington and released a statement calling for a revision of the nation's priorities to reduce war, poverty, and hunger on Earth. The Kennedy family's unofficial representative was the senator's brother-in-law, Sargent Shriver, the American ambassador to France. Shriver told the *Miami Herald* that he could vividly recall President Kennedy in his rocking chair on the day he addressed the joint session of Congress. "He told us, 'I firmly expect this commitment to be kept. And if I die before it is, all of you here now just remember when it happens I will be sitting up there in heaven in a rocking chair just like this one, and I'll have a better view of it than anybody.'" To NASA's relief, Shriver's comments garnered more attention from the wire services than the statement from his brother-in-law.

Less than an hour after the launch, Vice President Spiro Agnew arrived at the CBS News press facility for a live interview with Walter Cronkite. Since his selection as Nixon's vice presidential nominee the previous year, the fifty-year-old Maryland governor had done little to endear himself to the American public, the press, or his running mate. A number of gaffes on the campaign trail—including reports of ethnic slurs and callous comments—had defined him in the eyes of many as a national embarrassment.

Tom Paine had thought he'd found NASA's savior in the new vice president, who, following tradition, served as the head of the National Space Council. Seated next to Cronkite at his CBS desk, Agnew was asked about America's future in space. "I don't think we'd be out of line in saying we are going to put a man on Mars by the end of the century," the vice president announced. "I think the people in the country, the average man, want something to look forward to as an exciting objective."

Agnew repeated his Mars commitment to the cheers of NASA engineers in the Launch Operations Center a few minutes later. In reality,

Agnew had no influence in the White House, and his Mars statement had not been approved in advance. Newspaper editorials roundly criticized Agnew's infatuation with the Red Planet, and a scathing political cartoon published in *The Washington Star* depicted the vacantly grinning vice president as a space cadet proclaiming, "Mars or Bust." The cartoon was close to how the White House viewed him as well. Nixon's adviser John Ehrlichman sternly told Agnew to never again make any such comment about a Mars program.

Cronkite immediately followed his Agnew interview by welcoming Lyndon Johnson to his desk. The two politicians had arrived at the CBS facility simultaneously, causing a moment of panic among the producers, with Agnew receiving preferential treatment. Johnson was noticeably at ease as he settled into the seat. He soon attempted to place the legacy of Apollo in context, while alluding to Ralph Abernathy's concerns—though not mentioning him or the Southern Christian Leadership Conference by name. Why couldn't the cooperative effort that sent Apollo 11 to the Moon be directed toward fighting poverty, hunger, disease, and eliminating war, Johnson asked. "We must apply some of the great talent that we've applied to space to these problems." Cronkite wondered whether the principles of management pioneered by James Webb could be used to solve the nation's many other problems. In response, Johnson praised the man more than the theory. He asserted that Kennedy had hired the best possible person to oversee the job, defined the objective, and then given Webb the necessary resources and independence to get the job done.

Outside the small CBS facility, Arthur C. Clarke had watched the liftoff from the slightly elevated ground at the press center. Like Wernher von Braun, Clarke had devoted much of his adult life to making this day a reality. Remarkably, the geosynchronous-satellite network that Clarke had envisioned in 1945 had also just come to fruition. Intelsat III F-3, one of a new generation of satellites, was moved into position over the Indian Ocean a few days before Apollo 11's launch, thus completing the final link in the world's first geostationary-satellite network. Now also internationally recognized as the co-author of the most successful movie released the previous year, Stanley Kubrick's

2001: A Space Odyssey, Clarke had been signed by CBS to provide viewers with a vision of humanity's destiny in space.

The promise of the moment energized Clarke and Cronkite to look forward to what was to come. Clarke assured CBS's viewers that a future much like the one depicted in *2001* was on the horizon. Within the next five to ten years, crewed space stations would orbit the Earth, and the first permanent bases on the Moon would be established. Clarke predicted human flights to Mars and around Venus within twenty years. Much would depend on the development of nuclear rockets and reusable space vehicles to make this a reality. A few years from now, Clarke believed, Apollo 11's manner of going to the Moon would seem incredibly wasteful, comparing it to sending a huge ocean liner on a single voyage to transport three passengers and then sending the ship to the scrap yard.

CBS's Apollo 11 coverage was in fact heavily influenced by the optimistic vision of the spacefaring future depicted in Clarke and Kubrick's cinematic epic. After attending a screening of *2001* the previous year, Joel Banow, the CBS News television director assigned to produce its space coverage, began planning how to incorporate the excitement he felt in the movie theater into the upcoming forty-five hours of Apollo 11 special programming. He hired Douglas Trumbull, one of Kubrick's special-effects wizards, to create a special "guest expert" for the broadcast, a HAL supercomputer much like the one in the film, albeit without a psychopathic programming flaw. "HAL10,000" engaged with Cronkite silently via a printed display screen. It was a Hollywood special-effects illusion unlike anything seen on live television before, especially as part of a news broadcast.

Whenever listening to Clarke's unwavering rational optimism during their many on-air discussions, Cronkite appeared to be newly energized. Though much of the world was consumed in chaos, it was still possible for two middle-aged men to dream of new adventures, as if they were adolescents.

"Do you think that you and I will make a spaceflight?" Cronkite asked.

"I have every intention of going to the Moon before I die" was Clarke's reply.

"I hope you're right." Cronkite then added with infectious enthusiasm, "I'm dying to go."

THE STORY OF the three "amiable strangers" journeying to the Moon occasionally challenged journalists to add some element of suspense, color, or humor to their reporting. Luna 15 provided a little intrigue but no human drama. The astronauts' biographies were respectfully detailed, but by far the most engaging way to tell the story to watchers around the world was via the live color-television transmissions from the spacecraft. Julian Scheer had fought to include the cameras, and during the event they proved to be essential. The result was a primitive progenitor of what later became reality television: It presented the astronauts at work in their fascinating zero-gravity environment, somewhat self-conscious of being watched but very much unrehearsed and unscripted. It also helped to diminish any lingering cynicism about government crafting of the astronauts' public images.

In the six months since Apollo 8, the relevance of the space race between the superpowers had receded, replaced by the reality that the United States would soon meet its objective. Some within NASA, including Frank Borman and Bill Anders, believed that once a successful human moon landing had been accomplished, there was little practical reason to return. The United States would have proven its preeminence in space. Scientific lunar exploration and establishing moon colonies could wait. Space visionaries like von Braun, with ambitious goals for the near future, were never a faction with great persuasive power in Washington, except when there appeared to be a national threat. On the airwaves, however, the television networks were selling a thrilling adventure with great promise for tomorrow, an intoxicating extended broadcast for a world that had seen much bad news recently.

Four days after liftoff, on the Sunday when humanity's first attempt to land on the Moon was scheduled, the three commercial American television networks commenced live coverage of what promised to be a broadcast lasting more than twenty-four hours. CBS began its marathon in the late morning, with a prologue that featured correspondent

Charles Kuralt reading the opening lines of Genesis. At precisely the same moment, in the East Room of the White House, Frank Borman was reading the same lines, for a worship service at which President Nixon and more than 340 diplomats, congressmen, and government leaders were present.

In Houston's Mission Control room, flight director Gene Kranz and his White Team had taken over from the Black Team, headed by Glynn Lunney. At thirty-five, former jet-fighter pilot Kranz was one of the older men in the room. On this day there would be no live-television cameras broadcasting images from the room; however, a small team of NASA cinematographers would be recording the event for history.

Nor would there be live television from the Moon during the landing phase. To tell the story visually, the television networks had to rely on their own full-size studio simulations, prepared animated films, and scale-model spacecraft, supplemented by the live audio of the astronauts' voices. At CBS's New York Broadcast Center on West 57th Street, Walter Cronkite anchored the coverage from an elevated desk positioned in front of an artist's rendition of the Milky Way. The network's largest sound stage, which usually accommodated two of the network's afternoon soap operas, had been converted into "Space Headquarters" for the two-day broadcast. Cronkite had been so excited prior to the broadcast that he had slept little the previous night. In contrast, each of the Apollo 11 crew members had managed between five and six hours of sleep before their early-hour wake-up call.

By late morning of Apollo 11's fourth day, the crew had been in lunar orbit for nearly twenty-four hours. Armstrong and Aldrin had checked out all systems on the lunar module *Eagle* and were preparing to separate it from the *Columbia* command-and-service module. It was July 20.

Not long after Armstrong and Aldrin in the *Eagle* and Collins in *Columbia* reported that they had successfully undocked the two spacecraft disappeared into radio silence as their orbits took them around the Moon's far side. In Houston, those in Mission Control knew that one of three possible scenarios could play out within the next ninety minutes. The lunar module would either crash, abort the landing attempt, or successfully touch down on the lunar surface. Outwardly calm and confident, Gene Kranz addressed the White Team over a private audio channel. "The hopes and the dreams of the entire world are with us. This is our time and our place, and we will remember this day and what we will do here always." But no matter what happened, he assured them, he would stand by their decisions.

Eagle reemerged from the far side of the Moon at a height of only eighteen nautical miles above the surface, just minutes before beginning its powered descent. Collins in *Columbia* remained in a higher orbit, sixty miles above the surface. Despite months of advance planning, the landing sequence contained hundreds of unknowns that could compromise the mission. Amid the curt audio communications peppered with technical acronyms, Armstrong and Aldrin offered few outward signs that revealed what they were feeling. Their voices remained alert and focused. Some broadcasters occasionally felt the need to step back and absorb the moment with awe. "Just four and half minutes left in this era . . ." Cronkite interjected as the landing approached.

Suddenly, in the midst of the data-filled transmissions, both Armstrong and Aldrin called attention to a computer-program alarm unknown to them. There was a sudden atypical urgency in Armstrong's voice as he mentioned it and then followed with a request to Mission Control for an explanation. "Give us a reading on the 1202 Program Alarm." Armstrong received no clarification about the alarm from capcom Charlie Duke other than the words "We're go on that alarm." A few minutes later, when a related alarm sounded, Duke reassured Armstrong, "We're go. Same type. We're go. *Eagle,* looking great."

It was a moment of added suspense that no one had expected. Had the alarm occurred during a practice simulation on the ground, a landing abort would likely have resulted. But the first powered descent to the lunar surface 239,000 miles from home was a rather different situation. Armstrong was determined to land successfully, and every bit of his concentration was focused on piloting the lunar module. He intrin-

The Apollo command-and-service module photographed in lunar orbit. While orbiting the far side of the Moon and unable to communicate with Earth or with Eagle, *Michael Collins could speak to no one with the exception of his tape recorder. A NASA spokesperson observed, "Not since Adam has any human known such solitude." This unusual view of the spacecraft was taken during the flight of Apollo 16.*

sically trusted his reflexes during an emergency and knew this situation wasn't as serious as either Gemini 8's violent tumble or the lunar-training vehicle when it began swerving out of control. An esoteric computer program alarm was not about to keep him from the ultimate goal of his career.

His instincts proved right. The alarm had been caused by an unintended information-overflow problem in the lunar module's guidance system. Armstrong's attention soon turned to a bigger concern: The available fuel for the descent engine was running low as *Eagle* approached the Moon's surface. He had a good idea of his height as he came in for the landing, but he needed to find a safe location. There were more large craters than he'd expected, and he didn't want to maneuver the lunar module backward. With nothing like a rearview mirror, it was impossible to tell if he might place the back landing leg on

The lunar module Eagle *shortly after separating from* Columbia. *The LM was designed to function exclusively in space and was incapable of returning to Earth. Apollo 11 was only the third time the LM had been flown by astronauts during a space mission.*

an unstable crater rim. His view of the surface was also becoming obscured as a mist of lunar-dust particles was agitated by the descent engine's exhaust plume.

Armstrong hovered above the Moon's surface while gently maneuvering the lunar module to find the best landing spot. More than a half minute passed beyond the projected touchdown time. The descent engine's fuel was running dangerously low, with only a few seconds of propellant remaining. In Houston there was still no word whether they had landed.

"Contact light," Aldrin announced. A moment later Armstrong responded, "Shutdown."

Armstrong later said he neither saw the contact light illuminate nor heard Aldrin announce it. The landing was so smooth, Armstrong couldn't even tell the precise moment when they were on the surface. This was the climax of the mission, in Armstrong's mind. The first step and moonwalk to come a few hours later were secondary.

Then from the Moon came the official word as Armstrong announced, "Houston, Tranquility Base here. The *Eagle* has landed."

· · · ·

IT BEING A summer Sunday afternoon in the United States, not every American was watching television. Sunbathers at swimming pools and beaches listened to the voices from Houston and the Moon on portable transistor radios. Eight major-league baseball games were in progress when the *Eagle* landed at 4:17 P.M. Eastern Daylight Time. There was a brief interruption in the play to announce the news, followed by sustained cheering. In many ball parks, a performance of either "God Bless America" or "America the Beautiful" was played before the game was resumed.

The original flight plan called for Armstrong and Aldrin to rest before emerging from the lunar module at roughly 2:20 A.M. Eastern time on the early morning of Monday, July 21. However, before the flight there had been talk that the crew might be given an option to proceed onto the lunar surface five hours earlier. And about ninety minutes after landing, Armstrong indeed sent word that he would prefer to exit the lunar module at roughly 9:00 P.M. EDT.

Some Americans made a point of saying that they wouldn't bother to watch any coverage of Apollo 11. Lewis Mumford, one of the country's leading intellectuals, described the Apollo program as "a symbolic act of war [by a] megatechnic power system, in the lethal grip of the 'myth of the machine.'" Only a few minutes after the *Eagle* had touched down, Walter Cronkite wondered aloud how young Americans dismissive of the space program were reacting to the news. "How can anybody turn off from a world like this?"

Secretly, most of those who had expressed their disdain among their friends were also affected by the magnitude of the accomplishment. If one was human and curious, it was impossible not to be impressed and touched with a sense of awe.

Armstrong's innate piloting skill during the final minutes of the landing had provided a convincing argument for a human presence on crucial space missions. Eric Sevareid, a veteran CBS News journalist and commentator, told Cronkite, "As an old-fashioned humanist, it seemed to me a little reassuring that in those last seconds the human hand and eye had to take over from the computers . . . if it had all been computerized to the last second, we might have wrecked this *Santa Maria* on a rock."

Sevareid's comment proved to be uncannily apt. Russia's Luna 15, which the world press had been following for days, was secretly being readied for its automated lunar descent as Sevareid spoke. However, during Luna 15's final approach, it was unable to be slowed sufficiently to make a soft landing, and its telemetry transmissions halted abruptly at the moment it made impact with the side of a mountain. From the Soviet Union came only a brief announcement that Luna 15's mission had ended.

ALTHOUGH THE ASTRONAUTS deemed the moonwalk of secondary importance, the intense public curiosity surrounding Armstrong's first words and the promise of witnessing the historic moment as it happened on live television transformed the first steps on the lunar surface into a media event of greater magnitude than the landing itself.

How much advance thought Armstrong gave his first words remains open to conjecture half a century later. When questioned about it during the crew's press conference in Houston in early July, he grinned and merely said he hadn't given it any consideration. However, for a pilot routinely used to checking out an aircraft before taking it into the air, such lack of preparation appeared out of character.

Unlike Borman, Armstrong had decided not to entrust the assignment to a professional wordsmith. In fact, one story indicates that, contrary to what he attested, he did do some advance preparation. Armstrong's brother, Dean, recalls that during a lull in a board game they were playing a few weeks before the launch, Neil handed his brother a piece of paper on which he had written a few words. He didn't explain the context, but none was necessary. Dean Armstrong sensed his brother wanted some feedback. He responded with one word: "Fabulous." In magazines and newspapers, Truman Capote, Vladimir Nabokov, Anne Sexton, Sun-Ra, Joseph Heller, Timothy Leary, Leonard Nimoy, William Safire, and scores of other noted writers and celebrities offered their suggestions, ranging from the heavy-handed and the banal to the flippant and corny.

Truly the whole world would be watching and listening.

IN NEW YORK'S Central Park, a crowd had gathered in the Sheep Meadow on Sunday night to watch the moonwalk broadcast on a large screen. Unfortunately, it was beginning to rain gently, and preparations for the EVA and depressurizing the lunar module's cabin took far longer than expected. Movement inside the small cabin while wearing the stiffened pressure suits and large backpack life-support units was proving cumbersome. Rather than 9:00 P.M., it was a little after 10:30 P.M. when the door of the lunar-module was pulled inward to open.

A few blocks away from Central Park, Arthur C. Clarke joined others in a viewing area of CBS's Studio 41, anxiously waiting to see if the small television camera stored in an equipment bay on the LM would begin transmitting pictures. When Clarke looked around the room, he was amused to see not every eye on the television—one of the studio grips was deeply studying a racing form, oblivious to what was about to happen. Meanwhile, Wernher von Braun and other NASA managers were gathered in an observation room not far from Mission Control. Former administrator James Webb was watching the event on the television in his Washington home.

Though it was approaching 11:00 P.M. on the East Coast, few were worried about getting sufficient sleep before the workday. President Nixon had proposed that Monday, July 21, be a National Day of Participation, allowing Americans a day off, and asked employers to comply. The early editions of Monday's The New York Times and New York Daily News were already on sale. The Times had photo-enlarged its biggest available typeface to create the even-larger ninety-six-point headline MEN LAND ON MOON.

In southeast Australia, the Parkes Observatory dish antenna began to receive spectral, shadowy images from a small Westinghouse television camera, which Armstrong had deployed by pulling a release mechanism as he inched feetfirst out of the lunar module. The television camera had been added to the lander very late in its development, and only after Julian Scheer had learned that the only broadcast camera scheduled to fly to the Moon was the one designed exclusively for the command module. Initially, Houston told Scheer his request to put a television camera on board the lunar module was impossible because the LM's entire electrical and communications system had already been approved and certified. Besides, he was told, the Hasselblad still photography and 16mm movie film brought back by the astronauts would be sufficient. Undaunted, Scheer marched into James Webb's office and said, "I need you to issue an order to Houston that they will put a television camera on the lunar module." Webb agreed and made the call, and within a few months the engineers had redesigned the systems to accommodate the Westinghouse camera.

The black-and-white picture broadcast on American television sets was somewhat degraded from what was received in Australia. To transmit the signal, a television camera in Australia was positioned to shoot a monitor at the observatory, which was then relayed via the Intelsat satellite above the Pacific to a California receiving station and then over cables to Houston, where it was relayed to the television networks and to the rest of the world. The process also involved converting the image through various international television broadcasting standards, which further reduced the clarity. Nevertheless, the small eight-pound camera worked as planned, providing the first images ever seen from another world.

The picture was ghostly, but Armstrong could be seen descending the ladder of the lunar module. From a position at the bottom of the lander's footpad, Armstrong provided an initial report. "The surface appears to be very, very fine grained . . . It's almost like a powder." And then, after an extended pause, Armstrong delivered the words for which he will be forever known: "That's one small step for man, one giant leap for mankind."

Unfortunately, that moment was surrounded with a little confusion when it was broadcast on live television. On CBS, Walter Cronkite wasn't certain what Armstrong had said. "'One small step for man,' but I didn't get the second phrase," he confessed. Over at NBC, things weren't much better. Frank McGee was anchoring the network's moonwalk coverage and could only offer, "I think he said, 'One small

More than a half billion people around the world watched the live television broadcast as Neil Armstrong set foot on the lunar surface. At the time the ghostly black-and-white television pictures from the Westinghouse camera seemed remarkable. Within three years these TV images were much better: they were transmitted in color, and taken with an improved camera that could be operated by a technician sitting at a desk in Houston.

step for man, but one giant leap for man' . . . I'm not certain I got that all right." Within a few minutes, both Cronkite and McGee were provided with the full quotation.

As with Borman on Apollo 8, Julian Scheer had no advance knowledge what Armstrong would say. He was somewhat surprised at Armstrong's words but admitted he shouldn't have been. He thought Armstrong's reputation for being inarticulate was undeserved and attributed it to his penchant for saying only what was on his mind. When Scheer later confessed his surprise to Armstrong, the astronaut just shrugged and said, "You're not the *only* person who can speak the English language."

Armstrong's training for that first step included one bizarre precaution for a wildly unlikely scenario. Some lunar scientists theorized that the Moon's surface might be combustible and that when Armstrong's boot came into contact with the lunar soil, it could set off a reaction. Everyone thought the chances of this occurring highly remote, but to be safe Armstrong was advised to touch the heel of his boot gently upon the lunar surface when extending his initial step. After taking a few more steps and bending his knees in the suit, Armstrong reported that he was experiencing no trouble walking or moving around. Adjusting to maneuvering his body in the alien environment took only a few minutes.

His initial tasks before Aldrin joined him were to obtain a soil sample and take some photographs. After the sample was placed in a pouch attached to his leg, Armstrong took a panorama of the barren, flat landscape. He reported that he found it starkly beautiful, not unlike the high desert of California. "It's different, but it's very pretty out here." Aldrin's reaction when he joined Armstrong on the lunar surface was "Magnificent desolation."

The harsh sunlight was coming from an angle fourteen degrees above the horizon, this being a lunar morning, which would last longer than an entire earth day. The relatively low angle of the sun also meant the lunar surface hadn't yet reached the maximum temperature of 260 degrees Fahrenheit (127 Celsius).

Armstrong and Aldrin's first ceremonial task was to remove a protective metal cover from the memorial plaque mounted on the front leg of the LM. The Committee on Symbolic Articles intended the plaque's message to convey to the world that the lunar landing was a human achievement rather than a national political act.

Julian Scheer had drafted the initial text, but White House speechwriter William Safire had insisted on adding "A.D." (anno Domini) after "July 1969" so as to subtly acknowledge God. Scheer consented and the plaque was manufactured. But the matter didn't end there. A few days later, at another White House meeting, an assistant to President Nixon requested that "under God" be inserted after "We came in peace."

Astonished, Scheer asked him, "What God? This is a universal thing. What about the people on Earth who do not worship our God— Buddhists, Muslims, and . . ."

"Damn it, Julian, the president is big on God!" he was told. "I'm telling you, the president will want God."

A White House memo with revised wording including "under God"

HERE MEN FROM THE PLANET EARTH
FIRST SET FOOT UPON THE MOON
JULY 1969, A. D.
WE CAME IN PEACE FOR ALL MANKIND

NEIL A. ARMSTRONG
ASTRONAUT

MICHAEL COLLINS
ASTRONAUT

EDWIN E. ALDRIN, JR.
ASTRONAUT

RICHARD NIXON
PRESIDENT, UNITED STATES OF AMERICA

and initialed by Richard Nixon was given to Scheer. But after leaving the White House, Scheer did nothing. If a question arose, he would explain that the earlier version of the plaque had already been installed on the lunar module. When Scheer told Paine what he had decided to do, his boss simply said, "I didn't hear that. . . ."

Undoubtedly, had "under God" been added, atheist Madalyn Murray O'Hair would have issued a press release the next day. The lawsuit she'd filed against NASA was still in the courts when Apollo 11 landed on the Moon. So sensitive had the matter become after the Genesis reading that the Apollo 11 crew attempted to avoid anything that might be interpreted as an overt religious observance. When, shortly after landing on the lunar surface, Aldrin had requested that "every person listening in . . . pause for a moment and contemplate the events of the past few hours and to give thanks in his or her own way," few were aware he was in the process of taking communion with a small bit of bread, a plastic vial of wine, and a silver chalice. However, by the next day the reverend at Aldrin's Presbyterian church had revealed to the press that he'd helped to arrange the Moon's first communion.

A second silent religious observance during the moonwalk may have occurred as well. Filmmaker Theo Kamecke spoke with Neil Armstrong's grandmother shortly after the moon landing. "Neil made me a promise," she told him. "He promised the first thing he would do when he stepped on the Moon was to say a prayer." Kamecke concluded that this explained Armstrong's unusually long pause when he stepped off the lunar module; he was saying a silent prayer "just between him and the universe." The filmmaker decided not to recount that conversation until more than four decades later, worrying that during the 1960s it might be distorted and cynically retold.

It was estimated that when Armstrong and Aldrin unveiled the plaque, one-fifth of the world's population—600 million people—was watching, seeing themselves represented as fellow passengers on spaceship Earth, undefined by borders, languages, or religion. In comparison with the Apollo photographs of the whole Earth seen from space, the Apollo plaque made only a brief impression. But its message was the same: The people of the planet were one species, unified by universal hopes and dreams and motivated by a desire to explore and learn.

As Armstrong and Aldrin stood on the barren lunar terrain, they knew that every other living person was now hundreds of thousands of miles away, except for Michael Collins, who regularly passed overheard in *Columbia*. Yet they could never escape a conscious awareness that they were being observed the entire time. After mounting the small Westinghouse camera on a tripod to offer a long shot of their working area in front of the lunar module, Aldrin decided to give viewers a spontaneous demonstration. He realized everyone was curious to know what walking in one-sixth gravity was like and how easily he could maneuver on the lunar surface while wearing his bulky pressure suit. He positioned himself in the camera's field of view to demonstrate a few moves, all the while providing a verbal commentary. First he walked at a regular pace, then he tried a kangaroo hop; he followed that with a maneuver in which he moved rapidly and then attempted to change direction by countering his momentum, putting his foot out to the side and making a cutting motion, which he likened to a football player dodging down a field. As he performed all these motions he found he had to compensate for the way in which his large life-support back-pack affected his center of gravity. Never forgetting the small television camera watching him, Aldrin made an effort to remain within the lens's field of view. The demonstration was entirely spontaneous on his part, and the earthbound viewers loved it. It turned out to be one of the highlights of the moonwalk.

Before Armstrong and Aldrin deployed a small array of scientific instruments, there were two other ceremonial duties to perform. The first was the erection of the American flag, which had been outfitted with a simple metallic support to fully display it in the windless lunar environment. The second was a live telephone call from President Richard Nixon in the White House. As Armstrong and Aldrin stood next to the American flag, the president engaged in what he described as "the most historic telephone call ever to be made from the White House" with a message of goodwill penned by Nixon's speechwriter William Safire.

Despite the presence of the American flag, there was once again a conscious effort to balance nationalism with a more global message. "As you talk to us from the Sea of Tranquility," Nixon said, "it inspires us to redouble our efforts to bring peace and tranquility to Earth. For one priceless moment, in the whole history of man, all the people on this Earth are truly one—one in their pride in what you have done, and one in our prayers that you will return safely to Earth." The White House had originally wanted the astronauts to remain standing at attention for another few minutes while "The Star-Spangled Banner" was performed. When he learned of this plan, Frank Borman forcefully advised the president against it, arguing that it would only waste precious time that could be better spent on scientific and exploratory activities.

Armstrong and Aldrin spent most of the next hour collecting lunar samples and deploying the scientific instruments. Near the end of their moonwalk, the astronauts performed another ceremony, which received little notice. When Frank Borman had been in the Soviet Union, he received two Soviet medals, one commemorating Yuri Gagarin and the other Vladimir Komarov. Just days before the liftoff, Borman had given the medals to Armstrong and Aldrin, who placed them on the lunar surface along with an Apollo 1 patch memorializing Gus Grissom, Ed White, and Roger Chaffee.

Much of the time, they were outside the television camera's field of view, prompting the majority of Americans to retire to bed long before the astronauts managed to transport the 47.7 pounds of lunar rock and soil samples into the module's ascent stage and return inside. In total, Armstrong spent more than two hours on the lunar surface, Aldrin about an hour and three-quarters.

SCIENCE-FICTION AUTHOR RAY Bradbury was walking alone through the streets of London as the sun began to rise. He had been up the entire night. The previous evening, he'd departed the hotel where he was vacationing with his wife and children to be interviewed on live television by British broadcaster David Frost, as part of a marathon ten-hour "Moon Party" on the nation's commercial network ITV. (The BBC's news and science reporters were broadcasting coverage closer to what American television viewers were seeing.) Bradbury stood by and watched as Frost introduced a series of musical performances by Engelbert Humperdinck, Lulu, and Cliff Richard. Concluding that the program was too frivolous, Bradbury located the producer and told him he was walking out. "That idiot in there has ruined the greatest night in the history of the world! Get me a cab and get me out of here."

Bradbury found a taxi and took it across London to another television studio, where he recorded a brief conversation with CBS correspondent Mike Wallace, which was transmitted via satellite and broadcast with Cronkite's introduction a few hours later. In less than six minutes, Bradbury placed the moon landing in the context of humanity's eternal questions about the existence of God, the human need for religion, our existential search for meaning, the elimination of war, and prolonging the life of the species. "I believe firmly, excitingly, that we are God himself coming awake in the universe," he said. Space travel was humanity's existential effort to discover its place in the universe and would eventually result in traveling out to the stars after our sun died, millions of years from then. Furthermore, he argued, space travel had now given a new purpose for living and would eventually result in the elimination of war.

"The rocket and the exploration of space [are a] . . . wonderful moral substitute for war," he said. Men and boys actually loved war, even though they pretended not to, he confided. Now humanity should band together to ensure the species' existence. "All of the universe doesn't care whether we exist or not, but *we* care if we exist. . . . This is the proper war to fight."

Joyful and in tears, Bradbury left the studio and spent the rest of the night walking through the city, arriving at his hotel at breakfast time. Passing a news dealer, he saw the first papers with front pages displaying pictures of the moonwalk. One paper trumpeted the headline ARMSTRONG WALKS ON THE MOON, while inside ran a much smaller headline: BRADBURY WALKS OUT ON FROST.

Reactions to the achievement of Apollo 11 were coming in from all over the globe. CBS invited into the studio novelist Kurt Vonnegut, journalist Gloria Steinem, and science-fiction writer Robert Heinlein. The BBC interviewed author Vladimir Nabokov, who had rented a television for the occasion and described his thrill at watching Armstrong and Aldrin dance with grace "to the tune of lunar gravity." He expressed surprise that English magazine journalists appeared to have "ignored the absolutely overwhelming excitement of the adventure, the strange sensual exhilaration of palpating those precious pebbles, of seeing our marbled globe in the black sky, of feeling along one's spine the shiver and wonder of it."

Novelist and philosopher Ayn Rand's hatred for big government and socialism didn't quell her enthusiasm for Apollo 11 as a great heroic and technological triumph, even though *The New York Times* pointedly observed that the American "moon program was as socialistic in its central direction and financing as its rival Soviet effort." Although Rand had attended the launch and been among the notables in the VIP viewing area, she said she would have preferred Apollo as a private free-enterprise initiative. "If the government deserves any credit for the space program," she grudgingly conceded, "it is only to the extent that it did not act as a government—i.e., did not use coercion in regard to its participants."

There were also a few apathetic voices in the mix. Reached in France, Pablo Picasso told *The New York Times*, "It means nothing to me. I have no opinion about it, and I don't care."

Due to the Apollo coverage, conflicts around the globe briefly disappeared from the front pages of the world's newspapers. During the moon-landing weekend, Egyptian and Israeli planes engaged in combat over the Suez Canal, with aircraft lost on both sides. On the East Coast of the United States, journalists were dispatched to the island of Martha's Vineyard to cover an unusual developing political story.

Senator Edward Kennedy had driven his car off a one-lane bridge and into a tidal channel on the east side of the island, in the town of Chappaquiddick. The accident resulted in the death of the car's passenger, Mary Jo Kopechne, who'd been a volunteer in Robert F. Kennedy's presidential campaign. Attempting to downplay the incident, Senator Mike Mansfield, the Senate's majority leader said, "It could have happened to anyone." However, journalists soon confronted Kennedy with a number of unanswered questions surrounding his actions and his failure to immediately report the accident to the police. His presidential prospects in 1972 were suddenly in serious doubt.

When news of the situation reached the Manned Spacecraft Center in Houston, Tom Paine couldn't avoid thinking of his tense luncheon with Kennedy three weeks earlier. It seemed incredibly ironic that NASA was now "flying cover" for the senator by keeping his personal scandal off the front pages.

In the Soviet Union, the newspaper front pages were in stark contrast to those of the rest of the world. *Pravda* accorded Apollo 11 a two-paragraph story on the bottom of page five, and by the time that *Izvestiya* was issued in the evening, the moon landing was given a little more space, along with an illustration. As it had since its break from Stalinist Russia in 1948, Marshal Tito's socialist Yugoslavia followed its own path. The state public television network allowed full live coverage of an Apollo mission for the first time. As CBS's Bill McLaughlin reported from Belgrade, "Yugoslavia in fact has adopted the three

American astronauts as its own heroes. . . . This morning, anyone who could get near a television set did not do much work."

SOME OF THE nervous suspense surrounding Apollo 11's mission had dissipated: *Eagle* had successfully touched down on the initial attempt, despite the computer alarms and last-minute maneuvers; the world had heard Armstrong's first words and seen a live image of a human walking on the surface of the Moon; and Luna 15's mission had come to an abrupt end. Now all attention was focused on the lunar module's powered ascent from the surface, a procedure that had never been attempted, even by an automated spacecraft. There existed no contingency backup should the lunar module's ascent engine fail to operate properly. If it didn't work, Michael Collins would be forced to return alone. At the *New York Daily News*, a typesetter had already prepared a special disaster edition of the paper if the astronauts were stranded on the Moon. It was a stark front page with a one-word headline: MA-ROONED! It would to go to press and be on the streets in minutes if news from Tranquility Base turned tragic.

The White House was ready for disaster as well. When discussing the president's Apollo 11 preparations with speechwriter William Safire a few weeks before launch, Frank Borman asked about what the White House had planned should the astronauts fail to return. Safire was startled, suddenly realizing that he hadn't considered that possibility, and drafted a memo titled "In the Event of Moon Disaster," with a suggested presidential statement. Safire gave the memo to the president's chief of staff, H. R. Haldeman, who showed it to no one. He folded it and placed it in his suit pocket, at the ready until Apollo 11 was safely home.

Fortunately, *Eagle*'s launch countdown proceeded as flawlessly as had the Saturn launch the previous week at the Cape. The ascent engine worked perfectly, and the subsequent rendezvous and docking with the command module in lunar orbit was almost routine. Once again there was no live television, of either the lunar liftoff or the rendezvous in lunar orbit. The live voices of the astronauts provided the drama, with the television networks supplementing the visual storytelling with animation or simulations.

All that remained was leaving lunar orbit and reentry. Eight months earlier, the idea of executing both of these procedures had been fraught with suspense. Now, compared to what had just been accomplished, they seemed business as usual.

FRANK BORMAN AND Tom Paine joined Richard Nixon on the bridge of the aircraft carrier USS *Hornet* to view the return and recovery of *Columbia*. All three had flown on Air Force One to Johnson Island, southwest of Hawaii, where they then took helicopters to the *Hornet*. This was the first leg of an extended round-the-world presidential trip to countries in the Far East and Eastern Europe, including Romania, which would be newsworthy as the first visit by an American president to a communist country since the start of the Cold War. Nixon was hoping the world's excitement about the moon landing might work to the advantage of the United States and strengthen international relations—a likely reason the White House had code-named the trip "Operation Moonglow."

This would be Nixon's first encounter with Armstrong, Aldrin, and Collins, but as with the thwarted presidential dinner on the eve of liftoff, medical precautions once again intruded. Only minutes after *Columbia's* successful splashdown, the trio was required to don biological isolation garments, and the men were then helicoptered aboard the *Hornet,* where they quickly entered a specially prepared sealed living area constructed from a converted Airstream trailer. This would serve as a temporary quarantine facility until their return to Houston where they would spend the remainder of a planned twenty-one day period of isolation. All these precautions stemmed from the remote possibility that the astronauts could return with an unknown contagion. Few scientists believed it likely, but no one who had ever read the conclusion of H. G. Wells's *War of the Worlds,* in which the invading Martians are killed by exposure to a common earthbound virus, wanted to be wrong.

President Richard Nixon watches the recovery of Columbia *from the bridge of the aircraft carrier USS* Hornet. *With him are astronaut Frank Borman (right) and Admiral John S. McCain, Jr., Commander-in-Chief, Pacific Command. At that moment, Admiral McCain's son was confined in a North Vietnamese prison. Years later, after becoming Arizona's senior senator, John McCain III would reveal how the news that America had put a man on the Moon helped raise his morale during his lengthy captivity.*

So when Nixon finally came face-to-face with the Apollo 11 crew on live television, they looked at one another through a small plate-glass window and relied on microphones and speakers. Standing outside the trailer on the *Hornet*'s hangar deck, Nixon awkwardly engaged in small talk about baseball's All-Star Game before pronouncing the recent events "the greatest week in the history of the world since the creation." (President Nixon's friend the Reverend Billy Graham later cautioned the president that the claim may have been excessive.) Nixon's lunar telephone call had already come under criticism as an attempt to take credit for something undertaken by Kennedy and Johnson. An editorial writer likened what he saw as Nixon's appropriation of the historical occasion to a Khrushchev-like publicity stunt unsuited to an American president.

At this moment in his presidency, however, Nixon was just as intoxicated by the moon landing as was much of the world. The White

House correspondents noticed that, like Kennedy before him, Nixon appeared physically transformed whenever he was in the presence of the astronauts, often walking with a spring in his step, something they called the president's "moonwalk." As he'd gotten to know Frank Borman, Nixon began to regard the astronauts as the personification of everything great in the American spirit. "They are the sons he didn't have, the all-star teammates he never knew," a *Life* magazine writer observed. One White House staffer theorized that Nixon likely identified with their personal biographies. Much like Nixon, nearly all the astronauts had risen from modest beginnings and achieved greatness through a combination of discipline, courage, and a belief in God. He regarded them as products of the system, not rebels against it like the many student protesters on college campuses.

During their initial encounter on the *Hornet*, Nixon invited the crew to be honored with the largest presidential state dinner ever planned. The "Dinner of the Century," as the Los Angeles event a month later was nicknamed, was conceived as both a national patriotic observance and a television spectacular. The White House framed it as affirmation of traditional—meaning conservative—values, which they believed were under siege during a time of generational division, campus unrest, and revolutionary protest. The highly partisan guest list of 1,400 invitees reflected the polarized politics of the moment. Outside the Century Plaza Hotel on August 13, 1969, three thousand orderly antiwar protesters congregated as the diners arrived in limos. Inside, NASA officials uncomfortably dined with faded Hollywood celebrities and California Republican donors. Very few Democrats were invited, not even Alan Cranston, California's Democratic senator; Johnson's former vice president, Hubert Humphrey, was added at the last minute on the recommendation of Frank Borman.

Massive ticker-tape parades in New York and Chicago had greeted the astronauts on the same day that they flew to the West Coast for the president's Dinner of the Century. A long calendar of tributes, parades, formal dinners, and award ceremonies continued as the three men undertook an exhausting thirty-seven-day round-the-world "Giant Leap Goodwill Tour." The White House had granted the use of one of the president's 707s for an itinerary that took them to twenty-three countries and gave nearly one hundred million people an opportunity to see the first men to walk on the Moon.

The itinerary for the Giant Leap Tour included a visit to Zaire, but this was the only stop on the African continent. There had been no consideration that the crew would visit South Africa, a country that partnered with NASA to maintain a telemetry receiving station. No South African citizens had had an opportunity to watch the live televised moonwalk, due to the government's long opposition to allowing any television broadcasting; it was deemed a corrupting influence that would lead to crime, race-mixing, and communism. A small group of wealthy South Africans went so far as to charter a "Moon Tour" flight to London in order to see the broadcast live on television at the hotel Dorchester, but their compatriots could only rely on radio and newspaper accounts.

When the Cape Town *Sunday Times* reported that Michael Collins's wife, Pat, had been astounded to learn the reason why South Africa had not seen the moonwalk broadcast, her reaction induced a sense of shame in much of the country. Not long after, South Africans got a glimpse of what they had missed: Edited films of the moonwalk television broadcast were screened to segregated audiences at the Johannesburg Planetarium. After viewing "Moon Television"—as it was billed—South Africans began to realize that television wasn't the evil medium they had been informed for many years, and a vigorous national debate ensued, leading to the introduction of a national television service.

The citizens of the Soviet Union and China had been prevented from seeing the live lunar telecast as well, but for entirely different political reasons. When Russians heard the news of the moon landing, many reacted with disappointment not unlike the way Americans responded upon hearing of Gagarin's flight eight years earlier. In Moscow, Sergei Khrushchev, the son of the former Soviet premier, noticed that citizens, upon learning of Armstrong and Aldrin's accomplishment, tried to ignore the news; some were insulted. It seemed to him that most were not overly concerned, though, as they had their own day-to-day problems. Pro-Soviet news outlets derided Apollo as a wasteful extravagance that could have been accomplished much more cheaply with automated probes, and implied it was a stunt intended to distract from America's situation in Vietnam. On East German television, a news program informed its viewers that space stations and unmanned satellites were far more important; the Moon was a pointless destination and Apollo little more than a symptom of capitalism's failure. But, unlike some sensationalist American magazines that had made the case that the entire Soviet space program was, in fact, a clever propaganda hoax, there was no active Soviet disinformation campaign to spread rumors that the American government and media had faked the moon landing.

Within the privileged inner circles of the Soviet scientific world, 16mm copies of the Apollo moon-landing film were studied and discussed. And among those with access to the film was Sergei Khrushchev. For much of his career as an engineer, he had worked in the Soviet Union's missile-and-space program. He had personally witnessed his father's surprise when the news of Sputnik caused political shock waves in the West and had shared his country's pride on the day Vostok 1 was launched into space. In the early part of the decade it had seemed as if the Soviet Union was always one step ahead of the United States, even when some of the Soviet achievements—such as Leonov's perilous Voskhod 2 spacewalk—had been hastily improvised.

But following those early triumphs, much had changed. Concerned about the tremendous expense and hampered by infighting within the Soviet space program, Nikita Khrushchev didn't approve a plan to put a man on the Moon until more than three years had elapsed since Kennedy's moon-shot speech. The Russian lunar program didn't officially begin in earnest until after Khrushchev's removal in late 1964. Not long after, the untimely death of designer-manager Sergei Korolev came as an additional setback. In any case, the Soviets' decentralized management, the N-1 booster's fundamentally flawed design, and the expense of undertaking such an ambitious competitive program while lacking the financial, technical, and manufacturing resources equal to those of the United States practically doomed Russia's lunar quest from the beginning. The single reason for its existence was to best the United States, and once Apollo 11 had returned, the Russian program was quietly brought to an end without any acknowledgment that it had ever been under way.

In the course of a visit to his father's dacha in the late summer of 1969, Sergei brought along a copy of the Apollo 11 film. Together, father and son watched it with a mixture of emotions. Sergei felt both pride and envy—pride that human beings had actually achieved this astounding feat, and envy that America had done it first. His father was dismayed, unable to understand why the Soviet Union hadn't accomplished it but reluctant to cast blame or criticize those currently in power.

Quietly, at his dacha on the Black Sea, one of the two principal protagonists in the great space race that began eight years earlier watched it end.

ON THE AFTERNOON of July 20, CBS News correspondent Bill Plante and a camera crew found seventy-five thousand people gathering in the rain at New York City's Mount Morris Park. The crowd was not there to watch the moon landing but to see and hear Stevie Wonder, Gladys Knight and the Pips, and other Motown stars at the Harlem Cultural Festival. One attendee enthusiastically said he thought the moon landing was incredibly important. But, he then added, no more relevant in his life than the Harlem Cultural Festival. "I think they're equal."

Like many other American dailies, Chicago's leading African American newspaper, the *Chicago Defender,* ran a huge headline on the cover of its July 21 edition: WORLD STANDS STILL: MOON SHOT UNITES U.S. FOR INSTANT. The story on page three was remarkable for its singular opening line, which managed to be both celebratory and ironic: "The first non-racist moment in American history came at 3:17 P.M. [Chicago time] Sunday, when two Americans—nestled snugly in their lunar craft—became the first men to land on the Moon. At this moment, people of every race, nationality, age and condition were united in praise for an achievement symbolic of the American genius."

During his three weeks in quarantine, Neil Armstrong likely read accounts of the SCLC's protest at Cape Kennedy and may have seen

Nona Smith's letter in *The New York Times* confessing a difficulty identifying with those in the space program since not one black face was shown on television representing NASA. When the Apollo 11 crew landed in Los Angeles to dine with the president, a reporter present noticed Armstrong's reaction when he spotted a small African American boy kneeling among the cheering crowd. Armstrong deliberately pushed past scores of outstretched hands to reach the boy and shake his hand. Aldrin and Collins followed his lead, prompting thunderous applause from the crowd.

Since Ed Dwight's experience in 1963, there had been occasional appeals to try to integrate the astronaut corps. In the summer of 1967, the Air Force announced that Major Robert Henry Lawrence, Jr., had been selected as an astronaut for their military surveillance space program, which would use Gemini equipment. Lawrence's introduction was featured on the evening newscasts and on the front page of *The New York Times* under the headline NEGRO AMONG FOUR CHOSEN AS CREW OF MANNED ORBITING LABORATORY. Tragically, Lawrence died less than six months later, in a crash at Edwards Air Force Base. A trainee pilot flying the F-104 Starfighter lost control of the aircraft during a steep descent, and Lawrence was killed when his ejection seat fired sideways as the jet struck the ground. Had he survived, it is likely he would have been among the first astronauts to fly the space shuttle in the 1980s.

During the summer of Apollo 11, an integrated astronaut corps was even featured in American toy stores. No newspapers or television newscasts marked the occasion when Mattel introduced the black astronaut Jeff Long, a blue-space-suited action figure marketed as the "space buddy" of Major Matt Mason, Mattel's "Man in Space." In a Mattel television advertisement, Jeff Long drove a lunar tractor on the Moon, heading for Tranquility Base.

Geopolitics may have led to the creation of the lunar program, but culturally the emotional message of Apollo 11 reminded everyone of the determination conveyed by the words of President Kennedy's stirring Rice University address. On Broadway, the hit musical *Man of La Mancha* was in its third year, with its most popular song, "The Impossible Dream," embraced by audiences as a late-1960s anthem of optimism. Apollo demonstrated to the world that not only was it possible "to dream the impossible dream" but that it was achievable. While the world of the late 1960s was rife with turmoil and division, a better future seemed certain. We would be able to solve our problems and differences. Change was under way. We were smart enough to leave the planet and walk on another world. We would now be able to take on other projects here at home as well, the challenges that President Kennedy alluded to when he said, "We choose to go to the Moon in this decade and do the other things not because they are easy but because they are hard."

Much as David Lasser had predicted in *The Conquest of Space*, the first missions to the Moon contributed to a philosophical shift back on Earth, though it was hardly the immediate dramatic global transformation that he had optimistically forecast four decades earlier. The photographs of our fragile home planet accelerated an emerging environmental consciousness that a few visionaries had foreseen during the post-war era. But those images, along with personal reflections of the astronauts who traveled to the Moon, contributed to a second political and philosophical consciousness, which was given the name "the overview effect." Looking upon the Earth from a distance of nearly 240,000 miles and being able to obscure it from sight by raising a thumb prompted many of the astronauts to realize that the political, geographical, religious, and tribal differences that divided the peoples of the globe were glaringly trivial and dangerous for the future of the species.

For those who had put humans on the Moon, meeting Kennedy's mandate was only the beginning. Poppy Northcutt, who served on Mission Control's third shift during Apollo 11— Glynn Lunney's Black Team—had been elated as she watched the moonwalk on television. She, and the more than four hundred thousand other Americans whose work had led directly to that moment, knew that their efforts had made a lasting mark on the history of the human race. Julian Scheer thought that life on Earth had been changed for the better and as a result a new era of exploration had begun. Might the concerted energies and talents that had placed a human on the Moon now be applied to the nation's other priorities as well?

Arthur C. Clarke and Wernher von Braun looked forward to a new age of pioneers inhabiting colonies on the Moon and exploring Mars before the end of the century. Von Braun confidently predicted that a human would be born on another world by the year 2000.

Apollo 12 would visit the Moon in November, the next in what were to be nine more Apollo lunar expeditions, perhaps, many believed, to be followed in the early 1970s by extended stays of two weeks of more.

The possibilities ahead appeared practically limitless. The planets were waiting.

CHAPTER SEVEN

THE FINAL FRONTIER

(1970–1979)

PAN AM AIRWAYS was taking reservations for passenger flights to the Moon. There were no scheduled departure dates, the price was undetermined, and placing a reservation didn't require a down payment. But by the summer of 1969 they had issued more than twenty-five thousand personalized and numbered First Moon Flights Club cards to interested customers. Pictured on the back of the card was the space shuttle from Stanley Kubrick's film 2001: *A Space Odyssey*, which in the film displayed a prominent Pan Am logo on the outside of the passenger compartment.

California governor Ronald Reagan, Senator Barry Goldwater, and Walter Cronkite were all reported to be among the card-carrying members of the First Moon Flights Club. For many Americans, there remained little doubt that the future depicted in Kubrick's film would happen in their lifetime. Humanity had just put its foot on another world. It seemed assured there would be human settlements on the Moon and Mars before the end of the century.

Yet just a year after Apollo 11, there was little such talk in the White House. Vice President Agnew no longer spoke of going to Mars. Instead, he gained new prominence by delivering alliterative attacks on the administration's critics in the media and on college campuses. In his role as the administration's hatchet man, he derided political pundits and academics as "nattering nabobs of negativism" and "the hopeless, hysterical hypochondriacs of history." The vice president's performance as the voice of the conservative "silent majority" soon overshadowed his position as the National Space Council's chairman. Agnew's brief infatuation with the Red Planet was soon forgotten.

Officially, the White House continued to assert that President Nixon was a committed space activist who favored "a vital and forward-thrusting space program." But after the United States's third mission to land men on the Moon, the president's attitude toward the Apollo program had changed. Along with the rest of the world, President Nixon was caught up in the drama as Houston's Mission Control toiled over four suspenseful days in April 1970, attempting to bring the crew of Apollo 13 back to Earth safely after the command-and-service module was crippled by the explosion of an oxygen tank.

Despite the mission nearly ending in disaster, Apollo 13's safe return was celebrated as an astounding triumph of human ingenuity, bringing deserved attention to the resourceful teams of NASA flight controllers and the engineers in Mission Control. But it was a story that neither NASA nor the country wanted to dwell upon, and Apollo 13 was conventionally regarded as "the flight that failed" for the next quarter century—until it was retold for a new generation in commander Jim Lovell's bestselling memoir, which subsequently served as the basis of a popular Hollywood film.

For President Nixon, Apollo 13 was a traumatic turning point in his relationship with the space program, one that had been ironically foreshadowed nearly four months earlier when he invited the crew of the second Apollo moon-landing mission to stay overnight at the White House. During the visit, Nixon entertained the Apollo 12 crew and their wives in the White House screening room with a newly released Hollywood motion picture. *Marooned* was a stunning selection to screen on such an occasion: The space-disaster film, starring Gregory Peck, David Janssen, and Gene Hackman, was about the plight of three Apollo astronauts stranded in orbit as they slowly ran out of available oxygen. The wife of one Apollo astronaut later revealed that *Marooned* had given her nightmares.

During Apollo 13's ordeal, Nixon could do little more than helplessly watch a real-life version of the movie play out on national television. He closely followed NASA's efforts to bring the crew home but began worrying that if another accident occurred before the 1972 election, it

A Pan American World Airways "First Moon Flights" Club reservation card issued in 1969 featuring a picture of the Orion space shuttle from Stanley Kubrick's 2001: A Space Odyssey on the back. So successful was the Pan Am promotion that their rival, Trans World Airlines, began taking customer reservations as well, shortly after the flight of Apollo 8.

would reflect badly on his administration and undermine his diplomatic overtures to Russia and China. Nixon's changing attitude toward the space program even became apparent in the White House décor. The Apollo 8 "Earthrise" photograph, which had hung prominently near Nixon's desk in the Oval Office for the past year, was removed from display and was no longer to be seen.

NASA administrator Tom Paine had effectively won over Agnew during the early months of the administration. Despite the vice president's later reticence to campaign for grand space initiatives, Paine remained hopeful the Nixon White House would eventually come through with a bold and expansive commitment to NASA's future. With the successes of Apollo 8, Apollo 11 and the other four Apollo missions completed under his tenure as administrator, he believed the space agency had proven that it could deliver on its promise. In the months after Apollo 11, Paine went through rounds of frustrating negotiations with the White House's money men concerning NASA's upcoming budget, but with little resolution.

By 1970, NASA's share of the federal budget had fallen to 2 percent, less than half of what it had been five years earlier. The country was in a recession, and concern about rising inflation was casting a shadow over all future spending. Internally, the White House explored a number of scenarios for NASA's future, including a draconian option that would slash the agency's allotment even further, close both Houston's Manned Spacecraft Center and the Marshall Center in Huntsville, terminate all piloted missions after mid-1970, and concentrate almost exclusively on uncrewed landers to Mars and probes to the outer planets.

Most Americans assumed the space program consumed a far greater percentage of every tax dollar than was actually the case. Three months after Apollo 11 returned, an opinion poll indicated that 56 percent wanted Nixon to spend less money on space; only 10 percent spoke in favor of increasing NASA's budget. It was hardly the ideal moment for NASA to make the case for an all-out program to put humans on Mars by the early 1980s.

A few weeks before the launch of Apollo 11, Time Life Inc. had given Norman Mailer a contract to write a book about the first moon landing. Selections would be excerpted in *Life* magazine and the entire book published by the venerable Boston firm of Little, Brown. But when he submitted his finished manuscript in mid-1970—ten months past the deadline and more than twice its contracted length—Mailer and his publisher took stock of how much had changed in the past year. Little, Brown had just released *First on the Moon,* a first-person account by the Apollo 11 astronauts, ghostwritten by *Life* magazine staffers. The sales had been modest. In an understated internal memo written exactly a year to the day after Apollo 11 touched down, the publisher's sales manager noted, "I don't think that the moon landing is exactly the most commercial subject to write on these days," and suggested cutting the size of the proposed first printing of Mailer's *Of a Fire on the Moon* in half. Mailer was also mystified by the public's apathy and speculated about a possible marketing angle for his book. "Could we advertise the book by hitting hard on the fact that there was extraordinary interest in the moon shot just a little more than a year ago and now there is close to total indifference to the subject?" he wrote his editor.

But in the NASA offices, Paine remained undaunted. He asked the space agency's most persuasive spokesperson, Wernher von Braun, to leave his managerial position at the Marshall Space Flight Center and come to Washington to serve as NASA's deputy associate administrator for planning. After more than two decades in Alabama, von Braun decided to move his family to Alexandria, Virginia. However, not long after his arrival, he sensed a changed mood in the country and even wondered whether the lavish Apollo budget hadn't spoiled many of his colleagues. Reflecting on the current moment and his new assignment, von Braun confessed, "Too many people in NASA . . . are waiting for a miracle, just waiting for another man on a white horse to come and offer us another planet, like President Kennedy."

In an effort to better determine NASA's post-Apollo future, von Braun and Paine decided to hold a three-day off-site retreat where they would look ahead to the year 2000 and beyond. They chose as their venue a modest building on Wallops Island, one of the barrier islands on the Virginia coast, where NASA maintained a small but active research station and rocket-launching facility.

Prior to the gathering, Paine charged the twenty-five attendees to call upon their "swashbuckling buccaneering courage" and imagine the possibilities thirty years hence and all that NASA could accomplish by 2000. He cautioned his space-age band of brothers to "be careful of ideology, amateur social science, and economics," and open themselves to "a completely uninhibited flow of new ideas."

Von Braun asked his old friend Arthur C. Clarke to deliver the Wallops Island conference keynote address. The select group of attendees—not surprisingly, all men—included the directors of most of the NASA centers as well as another VIP, the most celebrated space traveler in the world, Neil Armstrong. It was the first time Clarke and Armstrong met each other, a moment that Clarke made certain was captured in a snapshot taken outside of the Wallops Station building.

The free-ranging discussions at the Wallops Island conference were more like something that might be overheard at a science-fiction convention than at a government-agency long-range-planning session. Moon colonies, giant space stations, and human settlements on Mars were projected by the year 2000. Also discussed were new types of nuclear-powered rocket engines that would facilitate human travel outside the solar system and the future of human evolution on other planets. When imagining life closer to home, attendees debated the feasibility of an intercontinental space plane, ways to free humanity from a dependence on agricultural-based food, and a combined global supercomputer and telecommunications network. The Wallops Station conference turned out to be the last hurrah of the grand visions seen in *Collier's,* at the New York World's Fair, and in the movie theaters where *2001: A Space Odyssey* was still playing more than two years after its world premiere.

Paine enthusiastically submitted his Wallops Island report to the White House, hoping that these daring ideas would find supporters in the administration and lead to future discussion. But Paine had also seen the writing on the wall. The Wallops Island report was his parting shot.

Within six weeks of submitting his conference summary, Paine tendered his resignation and returned to work at his former employer, General Electric. But before he left his NASA office, Paine steered future development a bit closer by canceling two Apollo lunar missions—Apollo 18 and 19—thus saving 40 million dollars from the NASA budget and allowing the design and development for a space shuttle to move forward. (Apollo 20 had already been canceled the previous January to allocate its Saturn V to the Skylab space station program slated for 1973.) Paine's decision was not popular. After the billions spent on Apollo, one astronomer expressed in frustration: "It's like buying a Rolls-Royce and then not using it because you claim you can't afford the gas."

In Houston, news of the two latest Apollo cancelations did not go over well. Astronaut Tom Stafford, in a rare break with decorum, spoke frankly to NBC News. He speculated that some astronauts might quit, saying the decision suggested the United States could become a second-rate power. But in the new age of limits, this was necessary triage. Not only had Paine kept the development of a future space shuttle alive, but the cuts to Apollo enabled the channeling of resources toward less costly yet hugely ambitious robotic missions to the outer planets.

For those hoping to spend New Year's Day 2000 at a hotel on the Moon, the dream ended in 1970. The following year, Pan Am quietly terminated its First Moon Flights Club promotion.

NINE YEARS LATER, the framed photograph of Arthur C. Clarke and Neil Armstrong taken at Wallops Station hung in a place of honor in the author's study at his home in Sri Lanka, the island nation off the southern tip of India that had removed all reference to its former British colonial name, Ceylon, in 1972. In the years since Armstrong and Clarke met, much had occurred, yet little followed the optimistic outline Clarke, von Braun, and Paine had envisioned at the June 1970 conference.

Not far away from the Wallops Island photograph on Clarke's wall of memories was a black-and-white image of him standing with von Braun. In 1972, in his new advocacy position in Washington, von Braun had helped NASA obtain presidential and congressional approval for the space shuttle, which had been scheduled to launch its first crew before the end of that decade. Von Braun had often spoken of hoping to be among its first passengers, though most of its crew members would come from a new class of thirty-five shuttle astronauts enlisted in 1978. Among them were the first women and people of color in NASA's history.

But by 1979 the first shuttle launch had been delayed into the early 1980s. Von Braun would never see it fly, much less ride on it. He died in 1977 at age sixty-five, shortly after leaving NASA. During his last years he served as vice president for engineering and development at

*Arthur C. Clarke and Neil Armstrong meet at the NASA long-term
planning session held at Wallops Island, Virginia, in June 1970. Although he had stepped on the Moon less than a year earlier, Armstrong attended the Wallops Island conference in a new role as a
NASA bureaucrat, the deputy associate administrator for aeronautics. He held his administrative position for only a year, resigning to
teach at the University of Cincinnati.*

Fairchild Industries, an aerospace firm. This had been the first private-
sector job in von Braun's career. No longer would he answer to dictators, generals, and government bureaucrats. But within months of
joining Fairchild, he was diagnosed with cancer and would die for years
later.

His death spared him further questions about his wartime past,
many details of which were about to come to public attention for the
first time. An investigative unit within the criminal division of the U.S.
Department of Justice, called the Office of Special Investigations, had
been created to delve into the histories of former Nazis who had entered the United States under fraudulent circumstances. Soon there

would be renewed interest in the Operation Paperclip Germans. In 1973, Arthur Rudolph, a key member of von Braun's Saturn V team, who had overseen the production of the V-2 at the Dora-Mittelbau slave-labor factory, was forced to renounce his American citizenship and return to live in Germany.

The past had been shadowing von Braun during his final years. He had been called to testify as a witness at a German war crimes trial shortly before the Apollo 11 landing. When he met with journalists and news cameras on the day of his testimony, von Braun declared that his conscience was clear and he had nothing to hide. Not long after, he appeared on TV as a guest on the late-night *Dick Cavett Show* to speak about NASA's future in space. But Cavett changed the discussion to ask a few pointed questions about his experience during the Third Reich, something that had never been raised during a television interview in the past two decades.

In the course of his short time at Fairchild, von Braun reconnected with his old friend Arthur Clarke, living in Sri Lanka's commercial capital of Colombo. Von Braun's initiative, in fact, eventually transformed Clarke's home in the elite palm-shaded Cinnamon Gardens neighborhood. When Fairchild's experimental ATS-6 satellite was positioned in a geosynchronous orbit over the Indian subcontinent in 1975, von Braun immediately thought of Clarke. ATS-6 was to be used in an international experiment to demonstrate the feasibility of beaming educational television programming to rural villages that couldn't otherwise receive signals. The Indian government would give a number of remote villages their own satellite ground stations, a receiver, and a television monitor from which the local community could gather to watch the broadcasts.

Von Braun suggested that one of the receiving stations be given to the man who had first proposed the geosynchronous-communications-satellite idea back in 1945. Not long after, a crew of six engineers from the Indian government arrived at Clarke's home, where they installed a sixteen-foot dish antenna on his second-floor patio-deck, transforming it into the only privately owned satellite-receiving station in the world. At the time Clarke also had the only working television set in the entire country; national TV broadcasting didn't come to Sri Lanka until 1979.

After nearly a quarter century as a frequent visitor to the United States, Clarke had chosen to reduce his busy travel schedule. Now, financially secure for the first time in his life due to his work on 2001: A *Space Odyssey,* the world came to him—both via satellite and whenever astronauts, authors, royalty, journalists, and other celebrities were passing through Colombo and wanted to say hello.

Over the years, Clarke had explained to journalists that he'd moved to Sri Lanka in 1956 for the skin diving. But as his friend the science writer Jeremy Bernstein later recounted, there were additional personal motivations behind his choice to relocate there. His decision came only a few years after the suicide of British mathematician Alan Turing, who had been convicted of gross indecency and given a choice between prison and chemical castration treatment as a supposed cure for homosexuality, then criminally outlawed in England. Clarke had been briefly married to an American woman in 1953, a decision he regretted soon after. Properly British and discreet, Clarke usually deflected inquiries about his personal life, occasionally making public comments that led listeners to infer he was heterosexual. But as he grew older he became less guarded. Reflecting on Turing's decision to commit suicide in 1954, Clarke confided to Bernstein "that if he'd had the chance he would have urged Turing to immigrate to the island."

Embedded in the tropical garden at the rear of Clarke's house was a small marker noting the grave of his beloved German shepherd, with a name and ancestry that commemorated the early space age:

SPUTNIK
1966–1978
SON OF LAIKA:
GENTLE, LOVING, FRIEND

Whenever he was showing off his house and garden and the satellite receiving station, Clarke told his guests that within a few years they would have access to a far more varied selection of television broadcasts from around the globe than he could currently access with his satellite dish. He explained that the world was still in a semaphore and smoke-signal era of communications when compared with what was to come. Soon telephones, radios, newspapers, and televisions would be combined into a single portable device with a high-definition screen and a keyboard, allowing for two-way communication with anyone on the planet. Such a device, which could be as small as a wristwatch, would allow anyone anywhere to access the world's great libraries. The integrated circuit that was used on the Apollo guidance computer a decade earlier had since been superseded by faster and vastly more sophisticated chips. In the United States the first personal computers were being sold to hobbyists, and Clarke was hoping to get one soon.

Though ever the optimist, Clarke was bothered by the recent resurgence of interest in paranormal phenomena and pseudoscientific fads in the United States and elsewhere. Books and television specials "investigating" ancient astronauts, Bigfoot, the Bermuda Triangle, and psychokinesis had become the rage. For decades Clarke had enjoyed listening to such theories and had maintained an open mind, but he now feared cranks were littering the fringes of science.

Perhaps the strangest and most baffling of the new fringe ideas in the mid-1970s was promoted in a self-published book arguing that the Apollo moon landings were all an elaborate hoax perpetrated by the American government. Bill Kaysing's *We Never Went to the Moon: America's Thirty Billion Dollar Swindle* had been written in the wake of the release of the Pentagon Papers, Watergate, and other disturbing government revelations. Motivated by anger about the Pentagon's deception in Vietnam, Kaysing decided to write something "outrageous," hoping that it might prompt Americans to no longer blindly accept as truth the official word out of Washington, D.C. Ironically, this was one rumor the Soviet Union hadn't tried to cultivate as part of their disinformation campaign against the United States.

At the time of the moon landing, few voiced rumors of it being a hoax, although as early as 1968 just such a secret government conspiracy had served as part of the plot of a darkly satiric BBC television drama, *The News-Benders*. Author Norman Mailer may have sensed the growing paranoia when he joked a year after Apollo 11, "In another couple of years there will be people arguing in bars about whether anyone even went to the Moon." He was quick to dismiss any such "mass hoodwinking" because, he argued, to pull it off would entail an effort and genius greater than the feat of launching the Saturn V and landing on the Moon. But as memories faded and disillusionment about established institutions became more widespread, this infectious idea slowly gained adherents.

By LATE 1979, America hadn't launched a single crewed space mission in four years. During the same time period, the Soviet Union launched fourteen flights carrying cosmonauts into earth orbit, many taking crews to their Salyut space stations. America's robotic planetary probes had taken over as the new stars of NASA. A pair of Viking landers transmitted the first color images from the surface of Mars in 1976, and three years later Voyager 1 began sending detailed images of Jupiter and its moons. It was followed by Voyager 2, which, like the first Voyager, traveled to Saturn, then, for the first time, headed out toward Uranus and Neptune. The Voyager program had originally been proposed in 1969 as a "Grand Tour" and would have included a visit to Pluto as well. But Congress's efforts to constrain the space agency's budget and NASA's desire to divert funding to the shuttle had reduced the Voyager missions' agenda.

Culturally, astronauts no longer held the iconic position that they had a few years earlier. *Life* magazine, which had published the astronauts' exclusive personal stories since 1959, ceased publication the same month as the final Apollo moon voyage, in late 1972. That same year a report revealed that NASA had disciplined the Apollo 15 crew members for smuggling a packet of unauthorized collectible autographed postal covers onto their flight, with the implied suggestion that they had an unethical plan to enrich themselves. Appearing not long after the release of the Pentagon Papers and other journalistic exposés, the Apollo 15 postal-cover scandal fostered the cynical assumption that everyone holding a position of privilege and power covertly cheated to get ahead—even America's most celebrated heroes. (In fact, the situation with the Apollo 15 crew was not entirely unique.)

Indeed, the next year the first book was published that dissected and debunked the carefully crafted images of the astronauts long promoted by NASA's public-affairs office and sustained by an adoring media.

By the end of the decade, the Apollo astronauts who had left NASA were appearing frequently in consumer advertising. Buzz Aldrin endorsed Volkswagen and its new computer diagnostics system. Frank Borman, now chief executive officer and chairman of the board of Eastern Airlines, became the public face of the company. Neil Armstrong, who had maintained a more private profile than most of his colleagues, surprised many by starring in an ad campaign for Chrysler at a time when the carmaker was experiencing serious financial trouble due to the energy crisis. Wanting there to be no lingering doubt about the identity of Chrysler's spokesperson, the advertising agency not only had Armstrong introduce himself on camera, but they also made sure his name was superimposed in large type below his face. The fading fame of the Apollo moonwalkers even prompted a successful television commercial with a balding man dressed in a checkered sports jacket, asking the camera, "Do you know me? I'm one of the astronauts that walked on the Moon. When I walk in here to rent a car, they don't always recognize me. That's why I carry an American Express card." Only later was the name of the third man to walk on the Moon, Apollo 12 commander Pete Conrad, revealed in a final close-up of his credit card.

More and more of the Apollo-era astronauts were joining the private sector, as the new class of shuttle astronauts was altering the popular conception of NASA's space voyagers. In public statements the generation of Mercury, Gemini, and Apollo astronauts, with years of test-pilot and combat-flying experience, acknowledged the transition as an inevitable sign of progress. John Young, who had previously entered space on four different missions, would be the only Gemini astronaut to fly the new space shuttle. But among themselves, the earlier generation regarded the new group of mission and payload specialists— astronauts assigned to oversee medical and engineering experiments or serve as civilian technical experts—as something different. One went so far as to refer to the new generation as "second-class astronauts."

However, before the first shuttle left the launchpad, something appeared that strongly reinforced the popular image of the astronauts from the dawning years of the American space program. It came first, in 1979, as an acclaimed book and later as a popular Hollywood movie. Together, Tom Wolfe's *The Right Stuff* and Philip Kaufman's screen adaptation became the dominant narrative of the early space age for generations of Americans.

Wolfe's book arrived while the country was still suffering from the aftereffects of the Vietnam War and the conflicted emotions that surrounded it. In the months before the book was published, depressing newspaper headlines told of an energy crisis, gas lines, a nuclear-plant mishap at Three Mile Island, and the president's concern about America's "crisis of confidence." *The Right Stuff* rescued the early jet test pilots and the graying Mercury, Gemini, and Apollo veterans from American Express commercials and elevated them as exemplars of a select, elite brotherhood, defined by the phrase that Wolfe appropriately chose as his book's title.

Elegiac and nostalgic, *The Right Stuff* aimed to celebrate something many believed had disappeared in the decade since Apollo 11: a precious, elemental kind of heroic American masculinity. "The right stuff" at its core was personified by the career of test pilot Chuck Yeager, a man whose career and influence had become so pervasive that Wolfe claimed that it could be heard in the cadence of every commercial-airline pilot's voice. A combination of rare skill and unwavering courage, "the right stuff" was what caused someone to willingly and repeatedly place his life on the line to fly an untested bit of machinery and push it to its limit. This was done humbly and without fanfare. The only accolades that mattered were those of their peers.

A leading practitioner of the New Journalism, Wolfe employed the tools of a novelist to reveal what he perceived as the essence of his story. His approach allowed him to examine the astronauts' complex psychology—their driving ambition, recklessness, courage, and flaws—with a freedom that had eluded the previous journalists who'd attempted to chronicle their story. The men portrayed in *The Right Stuff* are all recognizably human. They range from an excessively sanctimonious Boy Scout to a profane and bullying womanizer. One needs to relieve his bladder into his space suit during a delay on the launchpad; one drinks too much the night before breaking the sound barrier. Wolfe's high-achieving, brave alpha males live on the page in vibrant colorful vignettes that reveal how much was missing from the sanitized, colorless, and unrealistic portraits that appeared in *Life* magazine for more than a decade.

Yet Wolfe's liberties with narrative and approach were not without controversy. Colleagues of the late Gus Grissom took offense at how the book depicted his near drowning during the second Mercury mission and how the test-pilot elite at Edwards Air Force Base reacted with ridicule. More problematic was Wolfe's nostalgic celebration of a faded code of masculine meritocracy, seemingly in reaction to 1970s advances in gender and racial equality. *The Right Stuff* was written while the U.S. Supreme Court was considering the most decisive affirmative-action case of the decade, while the Equal Rights Amendment was awaiting state ratification, and as the first classes of women were attending the U.S. Air Force, Coast Guard, Military, and Naval academies. At the book's conclusion, Wolfe specifically signifies the twilight of the age of right stuff with Ed Dwight's experience at Edwards, presenting his story as a misguided early attempt at government-enforced affirmative action.

Released a year before the 1980 election of Ronald Reagan, *The Right Stuff* appealed to readers eager to remember an America when a president could inspire the nation to take on daunting challenges—even beating the Soviet Union in a race to the Moon. It called to mind a time before the political assassinations of the sixties, urban riots, wars in Southeast Asia, and the erosion of confidence in national institutions, yet it viewed the past through a skeptical lens informed by the culture of the seventies.

Perhaps due in part to its wistful evocation of faded American values, four decades after its publication, *The Right Stuff* remains the most influential narrative of the early space age. Its memorable set pieces deftly balance the courageous with the absurd, while never trivializing what Wolfe referred to as his "rich and fabulous terrain." The book's shadow falls across the hundreds of books and films that have

come since it was published. For those born after 1980, the space age
could be said to have begun on September 24, 1979, *The Right Stuff*'s
publication date.

THE ENDURING MEANING of the space race remains elusive half a century after it came to its end. Tom Wolfe's reexamination, written only a
decade after the moon landing, allowed him the opportunity to define
the rare combination of ambition, courage, and endurance that personified the first men to venture into the high frontier.

A great deal less celebrated than the early astronauts are the space
visionaries who looked into the future and whose youthful dreams instigated the race to the Moon. Fifty years after Apollo 11, the renowned
physicist Freeman Dyson, who knew both Arthur C. Clarke and Wernher von Braun, continues to actively contemplate our destiny as humans to migrate from the Earth into the cosmos. Though he played no
role in Project Apollo, in the 1950s Dyson was intimately involved with
the development of Project Orion, a research study that considered
the feasibility of a large spaceship powered by a series of controlled
nuclear explosions. A decade later, he briefly served as a consultant to
Stanley Kubrick during the filming of *2001: A Space Odyssey*. A space
visionary much in the spirit of von Braun and Clarke, Dyson now looks
back on Apollo as a marvelous achievement but one that was ultimately
a wasted opportunity.

While Kennedy's challenge helped channel government and public
support to meet a clearly defined decade-end goal, Dyson believes it
would have been far more productive if the first missions into space
had been delineated within the larger scope of space exploration over
the coming centuries. Had this been the case, Dyson says, entering
space would have been understood as merely an early step in humanity's ultimate destiny in the stars, somewhat in the manner of Russia's
mystical cosmists at the end of the nineteenth century.

Dyson likens space exploration to the human expansion across the
remote Polynesian islands of the Pacific and believes spaceflight should
be regarded as a series of steps in a quest—of men and women accept-

ing great risk to venture into the unknown. The spread of human settlements across the Pacific was not instigated by a pursuit of scientific
knowledge or as a form of political persuasion. What propelled that
exploration was an innate curiosity and the urge to move into new environments. "It was about taking chances—the essence of what makes
life interesting."

Though voiced by someone now in his tenth decade, Dyson's audacious vision of humanity's destiny in space may sound as startling to
twenty-first-century ears as those of Tsiolkovsky, Oberth, and Goddard
did a century ago. But his thoughts follow those of Clarke, von Braun,
and Tsiolkovsky, who metaphorically spoke of the Earth as merely the
cradle of the human race.

Dyson believes there is no more essential reason to send life into the
universe than to diversify the species. A need to live and function in
non-earthlike environments will lead to adaptations that will alter and
expand the branches of the human family. Rather than constantly
using a space suit, humans may develop a skin suitable to their sur-
roundings and lungs that will allow them to breathe freely in alien
habitats. Colonists on Mars may develop a furry exterior, as living comfortably on the Red Planet will be easier with an appropriate fur coat.

And as biotechnology becomes more sophisticated, Dyson predicts
this process of evolutionary adaptation will be engineered to happen
much faster than in the past. Human transformation might occur
within a few generations, possibly within one hundred years. The
scope of evolutionary change he foresees won't be confined to the
human species alone. Entirely new ecospheres could be engineered,
created to coexist in harmony with other creatures and flora developed
for that environment.

It's a vision very much from the pages of a science-fiction novel. In
the course of the Apollo program, the human species became the first
to walk, work, and even drive a car on an alien world. That journey to
the Moon began, in part, as a result of the mystical ideas of the Russian cosmists, who predicted the spiritual transformation of the human
race as it ventured into space. Freeman Dyson's vision also foresees
human transformation, albeit of a different sort, with biologically engi-

neered intelligent life diversifying the species as it colonizes other planets.

Humans first ventured into space little more than half a century ago, and within a decade they stood on the surface of another world. Those who walked on the Moon didn't regard themselves as the denizens of two worlds, but should transhumanism emerge as the inevitable legacy of space exploration, the moonwalkers themselves are likely to be considered its pioneers.

AFTER A FEW MORE
REVOLUTIONS AROUND THE SUN

BRIEF BIOGRAPHICAL NOTES ABOUT THE LATER LIVES OF SOME OF THE
PEOPLE WHO APPEARED IN THE PREVIOUS PAGES

When he addressed a joint session of Congress a few weeks after his return from his Apollo mission, BUZZ ALDRIN found it far more terrifying than landing on the Moon. Aldrin's difficulty coping with his daily life after becoming the second man to step on the lunar surface—his depression, anxiety, near-breakdown, and alcoholism—was recounted in a frank memoir, *Return to Earth*, published in 1973. Nevertheless, Aldrin became a vigorous advocate for a newly revitalized American space program, proposing an economical system with a series of spacecraft that would regularly cycle on scheduled journeys to Mars, back to Earth, and then out to Mars again on a continuing basis. He never flew in space again.

After serving as the backup command-module pilot for Apollo 11, BILL ANDERS moved to the nation's capital to serve as executive secretary of the National Space Council, where he witnessed the eccentric realities of Washington decision-making. He was surprised to observe the White House staff determine the space shuttle's final design based primarily on which prototype would deliver the greatest number of votes in the 1972 presidential election. After holding other government posts, including ambassador to Norway, Anders entered the private sector to take up management positions at General Electric, Textron, and General Dynamics, where he was chairman and chief executive officer. His one trip to the Moon transformed his own personal philosophical and spiritual outlook. "Here we are, on kind of a physically inconsequential planet, going around a not particularly significant star, going around a galaxy of billions of stars that's not a particularly significant galaxy—in a universe where there's billions and billions of galaxies. Are we really that special? I don't think so."

By the time he died at age eighty-two in 2012, NEIL ARMSTRONG had erroneously acquired a reputation as a recluse. Rather, Armstrong didn't enjoy giving interviews, having his life scrutinized, or leveraging his fame to enrich himself. He allowed for occasional public appearances but avoided being the center of attention and made an effort to share the limelight with others. Few of his personal items were sold during his lifetime, so when it was announced in 2015 that his widow had found a stowage bag of random items brought back from Apollo 11 in a household closet, it was treated as space history's equivalent of the discovery of King Tut's tomb. Fortunately, rather than being auctioned, the bag and its contents were loaned to the Smithsonian Air and Space Museum.

ABC News science correspondent JULES BERGMAN covered every space mission from the flight of Yuri Gagarin to the space shuttle *Challenger* disaster of 1986. Like Walter Cronkite, Bergman wanted to travel into space and informed his other colleagues in the press corps that he had already passed a rigorous exam similar to that which all the first NASA astronauts had endured. In the waning days of the Apollo era, Bergman was assigned to cover daredevil Evel Knievel's attempt to fly a rocket-powered Skycycle over Idaho's Snake River Canyon in 1974. Bergman was only age fifty-seven when he passed away in 1987.

Contrary to the wishes of some in the Nixon White House, FRANK BORMAN never entered politics. Shortly after leaving NASA, he joined Eastern Airlines, one of the nation's oldest and most prestigious air carriers, rising to the position of chief executive officer in 1976. After a few successful years, however, the effects of deregulation, debt, and squabbles with unions led to the sale of the airline and Borman's departure. He told his story in *Countdown,* published in 1988. He and SUSAN BORMAN observed the sixty-eighth anniversary of their wedding in 2018.

Precisely a year after the flight of Apollo 8, the California Chamber Symphony premiered a cantata composed by Jerry Goldsmith with a libretto written by RAY BRADBURY that reiterated his mystical thoughts on human space travel. By the 1970s, astrophysicist Carl Sagan and California's governor Jerry Brown echoed Bradbury's words predicting that space exploration would become the moral substitute for war. Shortly before his death in 2012, Bradbury recalled his CBS News Apollo 11 interview with Mike Wallace as among the high points in a long and illustrious life.

At his home in Sri Lanka, ARTHUR C. CLARKE connected to the world via the Internet, living out the vision of future communication he had predicted decades earlier. In the 1980s he reacquired his long-lost collection of American science-fiction pulp magazines, in the form of scanned digital files. Though post-polio syndrome confined him to a wheelchair during his final years, he maintained an active writing career until his death at age ninety in 2008. After the millennium, the multiplicity of moon-landing hoax/conspiracy theories on the Internet prompted Clarke to confess he didn't have time for lunatics: "I am too busy proving that George Washington never existed but was invented by the British Disinformation Service to account for a certain minor unpleasantness in the Colonies."

Subsequent to the Apollo 11 crew's Giant Leap Goodwill Tour, MICHAEL COLLINS accepted an offer from secretary of state William P. Rogers to serve as assistant secretary of state for public affairs. The experience was not pleasant and coincided with continuing American involvement in Vietnam, the incursion of American troops into Cambodia, and internal tensions between Rogers and national security adviser Henry Kissinger. Far more to his liking was his next assignment, as the director of the National Air and Space Museum, which opened its new exhibition building under his leadership in 1976. His memoir, *Carrying the Fire: An Astronaut's Journeys,* remains for many among the best firsthand accounts of the Apollo era.

Beginning a new life after leaving the Air Force in 1965, ED DWIGHT worked as an IBM sales executive and a construction entrepreneur, and in the 1970s he also earned an MFA in sculpture at the University of Denver. Not long afterward, he began a second career as an artist, specializing in retelling African American history in bronze, with depictions of the famous—Louis Armstrong, Charlie Parker, Hank Aaron—and the neglected—a scout from the battle of Little Bighorn, black cowboys, and an early Colorado settler. His memoir, *Soaring on the Wings of a Dream: The Untold Story of America's First Black Astronaut Candidate,* was published in 2009.

American comedian and activist **DICK GREGORY** often referenced the space program in his comedy of the 1960s, with lines like, "I heard we've got lots of black astronauts. Saving them for the first spaceflight to the sun." By 1968, however, he garnered additional press attention with a campaign as a write-in candidate for president of the United States. Later, he was among the first prominent celebrities to endorse a theory that the moon landings had been faked, part of an elaborate conspiracy between the two global superpowers, with the United States government trading wheat to the Soviet Union as hush money. He passed away in 2017 at age eighty-four.

DAVID LASSER retired from a management position at the International Union of Electrical, Radio and Machine Workers in 1968, the year after the Freedom of Information Act went into effect. He used a FOIA request to obtain the government's files on his background dating from the 1940s and discovered pages of false allegations and fabricated lists of political organizations that he never had joined. After years of trying to clear his name, Lasser received an official letter of public apology from President Jimmy Carter for his treatment during the Red Scare. In the years after the moon landing, he met and became friends with the man whose life had been profoundly influenced by *The Conquest of Space,* Arthur C. Clarke. He passed away in 1996 at age ninety-four.

WILLY LEY never found a permanent job in the aerospace industry, but his pioneering history of the early space age was first published as *Rockets* in 1944 and subsequently progressed through six different revised and expanded editions. The last, *Rockets, Missiles, and Men in Space,* was released in 1968 and praised as a seminal reference. Ley died suddenly less than a month before the launch of Apollo 11, at age sixty-two. In the week before the launch of Apollo 11, NASA received a letter proposing that some of Ley's ashes be among the items left on the Moon. While this request was never fulfilled, a crater on the far side of the Moon was named in his honor.

Following the Apollo 13 mission, JAMES LOVELL retired from the Navy and from NASA. Not long after, Hollywood attempted to tell the story of Apollo 13 in the form of a melodramatic TV movie focusing on the lives of fictitious characters at Mission Control, who, when not putting in long hours to bring the astronauts home, also have to deal with a bad marriage, a child-custody dispute, a dying father, and a dangerous heart condition. Lovell was so upset by the film, *Houston, We've Got a Problem*, that prior to the network broadcast he publicly complained that the producers had transformed one of the space program's finest hours into a trite soap opera. "I resent it," he wrote in a letter. "There's too much mixing up fact and fiction in the world these days." In 1994 he set the record straight with his book *Lost Moon*, which came to the screen as *Apollo 13*.

Named NASA's deputy administrator in 1969, GEORGE LOW was instrumental in the development of the space shuttle and moving the space agency toward Earth-based science. In 1976, Low left NASA to become president of Rensselaer Polytechnic Institute, and he was awarded the Presidential Medal of Freedom the day before his death from cancer in 1984 at the age of fifty-eight. His son David was named an astronaut in 1984 and flew into space on three space-shuttle missions.

After leaving the Kennedy administration and his position as chairman of the FCC in early 1963 to return to practice law in Chicago, NEWTON MINOW served as the co-chair of the Commission on Presidential Debates, where in 1976 he was instrumental in bringing about the first televised presidential debates since the Kennedy–Nixon campaign of 1960. He served as chairman of the Public Broadcasting Service and the Carnegie Corporation and in 2016 was awarded the Presidential Medal of Freedom.

For her work on the return of Apollo 13, POPPY NORTHCUTT was a recipient of the Presidential Medal of Freedom Team Award. After leaving the space program, she served on the board of directors of the National Organization for Women and was named by the mayor of Houston as the city's first women's advocate. She spent more than four decades on the front lines of the feminist movement, earning a law degree in 1984 and working as a prosecutor in the Harris County district attorney's office, and later had her own private practice specializing in issues of inclusion and reproductive rights.

HERMANN OBERTH traveled to the United States to witness the launches of Apollo 11 and, sixteen years later, the space shuttle *Challenger*. He remained interested in occult ideas and collaborated on books with a psychic who claimed she had received messages from extraterrestrials. More controversial, he became an idol of the German far right and briefly supported the radical National Democratic Party. During one of his visits to America, Arthur C. Clarke observed Oberth in a small group of visitors receiving a tour of NASA's Goddard Space Flight Center. Oberth was unrecognized by the young NASA scientists working there, who Clarke suspected were unlikely to even recognize his name. He lived to be ninety-five.

Following his departure from NASA in 1970, THOMAS PAINE returned to General Electric as a vice president and group manager and subsequently served as president and chief operating officer of the Northrop Corporation. In the 1980s he chaired the National Commission on Space for President Ronald Reagan, with an assignment to look ahead fifty years in order to establish space-program goals for the coming two decades. Paine had never been an enthusiast for the chosen design of America's space shuttle, which he likened to a "space truck with a mission no more glamorous than carting a load of toothpicks to Topeka." He died at age seventy, not long after serving on President George H. W. Bush's U.S. Space Policy Advisory Board.

The unofficial eighth astronaut of the Mercury program, COLONEL JOHN "SHORTY" POWERS spent the latter part of the 1960s endorsing commercial products, not only the long-running ad campaign for Oldsmobile but also Tareyton cigarettes and Carrier air conditioners. His syndicated "Space Talk" column appeared in hundreds of newspapers in the late 1960s, and his familiar cadence provided the voice-over narration for the

1966 Jerry Lewis space-themed sex comedy, *Way . . . Way Out*. After Apollo 11, Powers argued for diverting money from the Pentagon budget and giving it to NASA, reasoning that the technological advancements resulting from defense contracts could just as easily come from space research. When he died in 1979, he was only fifty-seven.

Less than a year after leaving NASA, astronaut **WALLY SCHIRRA** signed a reported hundred-thousand-dollar contract to appear with Walter Cronkite during CBS News's broadcast of Apollo 11. The network promoted the pairing as "Walter to Walter coverage," and they appeared together on space-related broadcasts through the Apollo-Soyuz Test Project in 1975. Schirra subsequently achieved additional media fame as a spokesperson for Actifed cold tablets, in a long-running series of television and print ads stemming from his battle with a head cold during Apollo 7. He authored an autobiography, *Schirra's Space*, published in 1988. He passed away in 2007 at age eighty-four.

Shortly after the Apollo 14 mission in 1971, **JULIAN SCHEER** was forced out of his position as head of NASA public affairs as a result of political pressure from the White House and some in Congress. He subsequently served as press secretary during North Carolina governor Terry Sanford's brief 1972 presidential campaign, as a Washington-based book scout for New York publishers, and as a vice president of LTV Corp. Over the decades he maintained a second career as an author of children's books, including *Rain Makes Applesauce*, a Caldecott honor book written while at NASA. He died in 2001 at age seventy-five.

Sidelined due to a heart condition prior to his scheduled 1962 Mercury flight, **DEKE SLAYTON** was finally cleared to fly into space a decade later. In 1975, at age fifty-one, he was one of three American crew members of the Apollo-Soyuz Test Project, the last flight of the Apollo command-and-service module, and the culmination of Richard Nixon's initiative to use space to achieve rapprochement with the Soviet Union. Slayton retired from NASA in 1980 and, two years later, as president of Space Services Inc. of America, became instrumental in successfully launching the first privately financed rocket into space. Slayton died in 1993 at age sixty-nine.

The only one of the original seven Mercury astronauts to walk on the Moon, ALAN SHEPARD returned to full-flight status in 1969, following surgery for an inner-ear disorder. As the commander of Apollo 14, Shepard, at age forty-seven, became at that time the oldest man to fly in space. Nothing he did during the entire mission garnered more attention than his brief attempt to hit two golf balls from the lunar surface. Though it stimulated new public interest in one of the later Apollo missions, Norman Mailer thought Shepard's golf stunt was in poor taste and dismissed it as a boorish celebration of locker-room culture. Shepard was one of the few men to have become wealthy while serving as an astronaut in the 1960s, largely as a result of his smart investments. He died in 1998 at age seventy-four.

By the early 1960s, the paper delivered in Barcelona by S. FRED SINGER about exploding an atomic device on the Moon had faded from attention. But it was far from the last provocative statement made by the noted atmospheric physicist during his long career. For more than four decades, Singer was quoted in national publications offering contrarian views on various policy issues, ranging from climate change ("global warming, if it occurs, is bound to be beneficial") to the dangers from secondhand cigarette smoke ("the risk for lung cancer . . . is not statistically significant"). In 2010, *Rolling Stone* dubbed him "the granddaddy of fake 'science' designed to debunk global warming."

After attending the launch of Apollo 11, JAMES WEBB watched the moon-landing coverage on television at his home in Washington. The former NASA administrator served for more than a decade on the executive committee of the Smithsonian Institution, as well as on corporate boards. He intentionally turned down positions on the boards of aerospace firms and refused lucrative offers to become a Washington lobbyist. A decade after his death in 1992 at the age of eighty-five, NASA renamed its Next Generation Space Telescope in Webb's memory, honoring him for his work fostering scientific research while running the space agency.

Historian Arthur M. Schlesinger, Jr., predicted that if the twentieth century was remembered for any single event five hundred years from now, it would be as the century when human beings began the exploration of space. "The generation that came of age in the 1960s [is] the last earthbound generation. They saw in their own lifetimes the shift of man as a creature of a single planet to man beginning the exploration of space. It's the most exciting and significant time in the history of mankind."

A half century after the first footsteps were impressed on another planetary body, less than a sixth of the world's population has a living memory of the event. Far fewer have a firsthand recollection of the circumstances that precipitated it. In not many decades, the events of the early space age—Sputnik, Vostok, Mercury, Gemini, Soyuz, and Apollo—will pass into a realm of history with no living witnesses walking the Earth.

To tell this story, we sought out participants, witnesses, and historians, who all provided hours of their time to record interviews used in the production of the six-hour *Chasing the Moon* documentary film and that also served as a foundation for this book. Conducted in private offices, homes, and conference rooms—one even took place in an airplane hangar—these extended audio discussions allowed for a casual intimacy that likely would have been impossible had film cameras been involved. We are indebted to Buzz Aldrin, George Alexander, Bill Anders, Valerie Anders, Joel Banow, Mark Bloom, Frank Borman, Ed Buckbee, Ed Dwight, Freeman Dyson, the late Theo Kamecke, Sergei Khrushchev, the late Jack King, Roger Launius, John Logsdon, Newton Minow, Poppy Nortcutt, and the late Leonard Reiffel for their generous willingness to tell their stories. Each meeting was an unforgettable encounter with living history.

Additionally, over the course of the four years during which the film and subsequently the book came into focus, we called upon the wisdom and assistance of many others who should be thanked: Tim Carey, Michael Chaiken, Marnie Cochran, David Cohen, Albie Davis,

Dwayne Day, Margaret Lazarus Dean, Paul Dickson, Eric Fenrich, Ian Greaves, Al Jackson, Boris Kachka, Tom Lehrer, Richard Lesher, Gideon Marcus, Neil McAleer, Larry McGlynn, Sarah Meltzoff, Leslie Morris, Steven Moss, Luke A. Nichter, James Oberg, Susan Scheer, Suzanne Scheer, Piet Schreuders, Roger Straus III, and Teasel Muir-Harmony.

We should single out our debt to the work of four scholars whose pioneering biographies of Arthur C. Clarke, Willy Ley, Wernher von Braun, and James Webb proved an invaluable resource to construct this written narrative. In particular, we urge any readers looking to learn more to seek out Jared Buss's "Willy Ley, The Science Writers, and the Popular Reenchantment of Science"; W. Henry Lambright's *Powering Apollo: James E. Webb of NASA*; Neil McAleer's *Odyssey: The Authorized Biography of Arthur C. Clarke*; and Michael J. Neufeld's *Von Braun: Dreamer of Space, Engineer of War*. Additional mention should be made of Eric Leif Davin's *Pioneers of Wonder*, with its invaluable chapter and interview with David Lasser; John Elder's paper "The Experience of Hermann Oberth"; and Asif A. Siddiqi's "Imagining the Cosmos: Utopians, Mystics, and the Popular Culture of Spaceflight in Revolutionary Russia."

This book would not exist if not for Scott-Martin Kosofsky and the creation of David Meerman Scott and Rich Jurek's *Marketing the Moon,* which somewhat serendipitously and circuitously led to this project's origin. Both David and Rich have been followers of this project since its inception, and throughout the writing of this book, Rich has offered invaluable insight and support, all while simultaneously working on the creation of his own pioneering biography of George Low.

The idea for this book was inspired by the initial research we did for what eventually became a PBS six-hour documentary miniseries of the same name. Producers Keith Haviland, Daniel Aegerter, and Ray Rothrock provided invaluable early support that allowed us to begin the research and to conduct the key interviews for the film. We are also incredibly indebted to executive producer Mark Samels, senior producer Susan Bellows, and the entire team at PBS/American Experience for making the film a reality.

These pages would not have been possible without considerable support provided by Mark Andres, Peggy Hogan, and Lucinda Jewell, whose lives have been shaped by a commitment to the importance of books, and, in the case of *Chasing the Moon*, generously assisted in its creation. Alan Andres wants to thank his partner, Anne Weaver, for both her belief in this project and her willingness to be primary family caretaker while it was being written. He is also very appreciative of the curiosity—and patience!—of Ian and Miranda as they listened to their father talk for what may have seemed to be an eternity about this history from a half century ago.

From our initial inquiry, Jane von Mehren had little doubt that contemporary readers would be intrigued by our approach. Working with her again after many decades was one of this project's additional rewards. Our editor at Ballantine, Susanna Porter, has demonstrated why she is so highly valued in the publishing world. Her enthusiasm, patience, astute questions, and careful eye have transformed these pages in profound ways of which the reader is unaware. Additionally, copy editor Kathy Lord's scrutiny, diplomacy, and wisdom are hidden within every paragraph of this book. Her diligence and attention to the smallest details are appreciated more than a small acknowledgment reveals. Thanks also to Emily Hartley, Deborah Foley, Matthew Martin, Simon Sullivan, Greg Kubie, and Dennis Ambrose for their work behind the scenes keeping everything on schedule.

Every effort has been made to trace the holders of all copyrighted material. We greatly regret any inadvertent omissions or errors within these pages; should they come to our attention, we will attempt to rectify them in future editions.

ACC: Arthur C. Clarke
NASM: National Air and Space Museum
NYT: *The New York Times*

CHAPTER ONE: A PLACE BEYOND THE SKY

4 **Now he learned** ACC, introduction to *The Conquest of Space* by David Lasser (Burlington, ON: Apogee Books, 2002).

6 **A full decade after Tsiolkovsky's groundbreaking paper** Michael J. Neufeld, "The Three Heroes of Spaceflight: The Rise of the Tsiolkovskii-Goddard-Oberth Interpretation and Its Current Validity," *Quest* 19, no. 4 (2012): p. 8.

7 **"As I looked"** Robert Goddard, *The Papers of Robert H. Goddard 1898–1924* (New York: McGraw-Hill, 1970), p. 9.

7 **He also proposed** "Invents New War Weapon: Armistice Prevented Use of Dr. Goddard's Rocket," *NYT* (March 30, 1919).

8 **Within weeks, *The New York Times*** "First Volunteer for Leap to Mars," *NYT* (February 5, 1920).

8 **The *Times* slammed Goddard** "Topics of the Times," *NYT* (January 12, 1920).

8 **Occasionally Goddard was complicit** "Plans Ocean Rocket Carrying Passengers; Professor Goddard Experiments on a New Method for Crossing the Atlantic at Terrific Speed," *NYT* (July 4, 1927).

8 **A brilliant student of mathematics** John Elder, "The Experience of Hermann Oberth," *History of Rocketry and Astronautics: Proceedings of the Twenty-Fifth History Symposium of the International Academy of Astronautics* (San Diego: American Astronautical Society, 1997), pp. 278–280.

9 **"This was nothing but a hobby"** Hermann Oberth, "Autobiography," in *The Coming of the Space Age*, ed. ACC (New York: Meredith, 1967), p. 116.

11 **One of Tsiolkovsky's** Frank H. Winter, *Prelude to the Space Age: The Rocket Societies, 1924–1940* (Washington: Smithsonian Institution Press, 1983), p. 26.

11 **In 1924, Russian** Asif A. Siddiqi, "Imagining the Cosmos: Utopians, Mystics, and the Popular Culture of Spaceflight in Revolutionary Russia," *Journal of Space Mission Architecture*, no. 1 (Fall 1999): p. 271.

12 **Lasser, the son of Russian immigrant** Eric Leif Davin, *Pioneers of Wonder: Conversations with the Founders of Science Fiction* (Amherst, NY: Prometheus Books, 1999), pp. 31, 46–47.

12 **Avid readers noticed** Lester Del Rey, *The World of Science Fiction, 1926–1976* (New York: Ballantine, 1979), p. 50.

12 **Goddard informed Lasser** Davin, *Pioneers*, p. 56.

13 **In one of his first roles** *Bulletin of the American Interplanetary Society* Vol. 1, no. 7 (1931); "2,000 Ride Rocket to Moon in Museum," *NYT* (January 28, 1931).

13 **Lasser had concluded** Davin, *Pioneers*, p. 53.

13 **"We learn that"** Lasser, *The Conquest of Space*, p. 17, 88.

14 **He soon amassed** Clarke's school notebooks are now in the collection of the Smithsonian's NASM.

14 **But when he read** ACC, "In the Beginning Was Jupiter," *NYT Book Review* (March 6, 1983).

14 **Archie Clarke's parents** Fred Clarke, *Arthur C. Clarke: A Life Remembered* (Burlington, ON: Apogee Books, 2013), p. 17.

15 **Like many other curious boys** Neil McAleer, *Odyssey: The Authorized Biography of Arthur C. Clarke* (London: Gollancz, 1992), p. 6.

15 **He loved reading** ACC, introduction to *The Collected Stories of Arthur C. Clarke* (New York: Tor Books, 2001), p. ix.

16 **All of the serious** Winter, *Prelude,* p. 42.

17 **One of the society's remaining** Winter, *Prelude,* pp. 44–48.

18 **Unemployment in the United States** Davin, *Pioneers,* p. 56.

18 **"If you like working"** Davin, *Pioneers,* p. 57.

19 **"I am extremely interested"** ACC, letter to L. J. Johnson in *Odyssey,* p. 11.

19 **On its final transatlantic voyage** "Here to Show Us How to Use Mail Rockets," *Brooklyn Daily Eagle* (February 21, 1935).

19 **Ley had even written** Jared S. Buss, "Willy Ley, the Science Writers, and the Popular Reenchantment of Science" (PhD diss., University of Oklahoma, 2014), pp. 114–116.

22 **"I prided myself"** ACC, *Astounding Days* (London: Gollancz, 1989), p. 125.

25 **When the project was completed** ACC, *Astounding,* p. 152.

25 **"When a distinguished"** Clarke's other two laws being: "The only way of discovering the limits of the possible is to venture a little way past them into the impossible" (Clarke's Second Law, 1962), and "Any sufficiently advanced technology is indistinguishable from magic" (Clarke's Third Law, 1973).

25 **News about the society's rocket** Tony Reichhardt, "H.M.S. Moon Rocket," *Air & Space* (March 1997).

25 **During the flurry** McAleer, *Odyssey,* p. 35.

26 **Sitting on a veranda** Fritz Lang, "Sci-Fi Film-maker's Debt to Rocket Man Willy Ley," *Los Angeles Times* (July 27, 1969).

28 **Laughter was heard** ACC, *Astounding,* pp. 153–154.

28 **Should Ley have needed** "V-2 Details Are Revealed," *Life* (December 25, 1944): pp 46–47.

30 **Ley published an article** Buss, "Willy Ley," pp. 232–233.

30 **"Those that are still alive"** Willy Ley, "V-2 Rocket Cargo Ship," *Astounding Science Fiction* (May 1945).

30 **In the mid-1930s** Michael G. Smith, *Rockets and Revolution: A Cultural History of Early Spaceflight* (Lincoln, NE: University of Nebraska Press, 2014), p. 285; Jay Robert Nash, *Spies: A Narrative Encyclopedia of Dirty Tricks and Double Dealing from Biblical Times to Today* (New York: M. Evans, 1997), p. 130.

30 **Technology for both rockets** Frank H. Winter, "Robert Goddard Was the Father

of American Rocketry. But Did He Have Much Impact?" AirSpaceMag.com (May 8, 2018).

31 **From an American military officer** Buss, "Willy Ley," pp. 259–261.

31 **"I only hope"** Willy Ley, letter to Robert Heinlein, "Willy Ley" (August 29, 1945), p. 261.

32 **"Entire German staff"** "Reich Experts Identified," *NYT* (November 18, 1945).

33 **"A crackpot with mental delusions"** Davin, *Pioneers*, p. 54.

33 **Lasser was incredulous** Lynn Darling, "A Rebel Redeemed," *The Washington Post* (April 23, 1980).

34 **Instead, the White House** Annie Jacobsen, *Operation Paperclip: The Secret Intelligence Program that Brought Nazi Scientists to America* (New York: Little, Brown, 2014), p. 195.

35 **Early in 1945** ACC, "Peacetime Uses for V-2," *Wireless World* (February 1945): p. 58.

35 **During that summer** ACC, "Extra-Terrestrial Relays: Can Rocket Stations Give World-Wide Radio Coverage?" *Wireless World* (October 1945): pp 305–308.

35 **A second, less historically important** McAleer, *Odyssey*, p. 65.

36 **"In a very short time"** ACC, *Ascent to Orbit: A Scientific Autobiography* (New York: Wiley, 1984), p. 83.

36 **Bonestell's scientifically accurate** Ron Miller, *The Art of Chesley Bonestell* (London: Paper Tiger, 2001), p. 36.

38 **In Rahway, New Jersey** Ray Spangenburg and Diane Moser, *Carl Sagan: A Biography* (Westport, CT: Greenwood, 2004), pp. 10–11; Carl Sagan, "In Praise of Arthur C. Clarke," the Seth MacFarlane Collection of the Carl Sagan and Ann Druyan Archive, Library of Congress (October 28, 1981).

CHAPTER TWO: THE MAN WHO SOLD THE MOON

39 **It's no surprise** *Camel News Caravan*, 16mm film segment, NBC Universal Archives, clip no. 5112458708 (March 13, 1952).

40 ***Collier's* readers were introduced** Cornelius Ryan, "Man Will Conquer Space Soon: What Are We Waiting For?" *Collier's* (March 22, 1952): pp. 23–24.

40 **This wasn't the first meeting** Jared S. Buss, "Willy Ley, the Science Writers, and the Popular Reenchantment of Science" (PhD. diss., University of Oklahoma, 2014), pp. 266–268.

40 **Physically, he could** Michael J. Neufeld, *Von Braun: Dreamer of Space, Engineer of War* (New York: Knopf, 2007), p. 43.

41 **He wore thick-lens eyeglasses** "Longines Chronoscope," kinescope, CBS-TV, National Archives and Records Administration (August 4, 1952), 200LW124.

41 **Over glasses of wine** Buss, "Willy Ley," pp. 266–268.

41 **Von Braun revealed** Here the discussion between Ley and von Braun on the evening of December 6, 1946, is used to tell von Braun's history from 1932 to 1945;

it also references other post-war accounts that von Braun gave about World War II.

42 "We hit the big time!" Wernher von Braun, interview in "Hitler's Secret Weapon," *Nova,* WGBH Boston (January 5, 1977).

42 As their conversation continued Buss, "Willy Ley," p. 269.

42 Von Braun revealed that "Hitler's Secret Weapon," *Nova.*

43 When the screening ended "Hitler's Secret Weapon," *Nova.*

45 Two days before von Braun Frederick Graham, "Nazi Scientists Aid Army on Research," *NYT* (December 4, 1946).

45 Someone unimpressed Buss, "Willy Ley," pp. 269–270; Robert Heinlein letter to Captain Cal Laning (January 12, 1947), in "Willy Ley," p. 268.

46 It was while quartered Neufeld, *Von Braun,* pp. 229–230.

47 Unfortunately for von Braun Neufeld, *Von Braun,* p. 246.

48 "Despite the grief" Wernher von Braun, letter to British Interplanetary Society, Daniel Lang, "A Reporter at Large: A Romantic Urge," *The New Yorker* (April 21, 1951).

48 During his brief return Neufeld, *Von Braun,* p. 233.

48 "It is a sad reflection" "Plan On Germans Scored; Admission of Scientists with Citizenship Offer Attacked," *NYT* (January 5, 1947).

49 But after seven decades Beverly S. Curry, *The People Who Lived on the Land that Is Now Redstone Arsenal* (Summerland Key, FL: B. S. Curry, 2006).

49 The German engineers who arrived Lang, "A Reporter at Large," *The New Yorker.*

50 Without much further discussion Buss, "Willy Ley," pp. 317–318.

50 Over cocktails, Ryan Wernher von Braun, draft of speech, in *Von Braun,* p. 256.

52 In April 1951 Lang, "A Reporter at Large," *The New Yorker.*

53 In the weeks before the March publication Neufeld, *Von Braun,* p. 258.

53 "Space rockets are" Wernher von Braun, letter to Cornelius Ryan, in *Von Braun,* p. 258.

55 The former Speaker of the House "Text of Chairman Joe Martin's Address to GOP Convention," *Arizona Republic* (July 10, 1952): p. 15.

55 Clarke sat up ACC, unpublished speech (July 1952), in *Odyssey: The Authorized Biography of Arthur C. Clarke* by Neil McAleer (London: Gollancz, 1992) p. 90.

57 The cover illustration Robert Heinlein, *The Green Hills of Earth* (New York: New American Library, 1952); Piet Schreuders, James Avati, and Kenneth Fulton, *The Paperback Art of James Avati* (Hampton Falls, NH: Donald M. Grant, 2005).

57 Heinlein confessed Robert Heinlein, letter to Sandra Jane Fulton, Robert A. and Virginia Heinlein Archives, UC Santa Cruz.

57 Writing in an essay ACC, "Sinbad in a Spaceship," *NYT Book Review* (November 16, 1952).

57 In a lengthy diversion Neufeld, *Von Braun,* pp. 284–285.

58 Disney's choice David R. Smith, "They're Following Our Script: Walt Disney's Trip to Tomorrowland," *Future* (May 1978), p. 57.

58 **"If we were to start"** Wernher von Braun, "Man in Space," *Disneyland* (March 9, 1955).

59 **Forty million** Willy Ley, "The How of Space Travel," *Galaxy Science Fiction* 11, no. 1 (October 1955): p. 60.

59 **Polling conducted** William Sims Bainbridge, *The Meaning and Value of Space-flight: Public Perceptions* (New York: Springer, 2015), p. 6.

60 **And in a final act** *Rocket Experts Become Americans*, American newsreel (April 14, 1955).

60 **The Disney studio** *A Walt Disney Science Factual Production: Man in Space*, Educational Film Division, Walt Disney Productions, Burbank, CA.

60 **Released in late 1955** *Challenge of Outer Space*, 16mm, Department of Defense (October 20, 1955).

62 **Von Braun believed** Neufeld, *Von Braun*, p. 298.

62 **"The statement would hurt"** Wernher von Braun, letter to Ward Kimball (August 30, 1955), Smith, "They're Following Our Script," p. 59.

63 **At Redstone Arsenal and in the city** Paul Dickson, *Sputnik: The Shock of the Century* (New York: Walker, 2001), p. 90.

63 **Four days after the White House's** John Hillaby, "Soviet Planning Early Satellite; Russian Expert in Denmark Says Success in 2 Years Is 'Quite Possible,'" *NYT* (August 3, 1955).

63 **Korolev distrusted the German engineers** Anatoly Zak, "The Rest of the Rocket Scientists," *Air & Space* (September 2003).

64 **It was approaching midnight** McAleer, *Odyssey*, p. 141.

64 **"As of Saturday"** ACC, "Free Zone Urged in Outer Space," *NYT* (October 10, 1957).

65 **Singer's paper was titled** S. F. Singer, "Interplanetary Ballistic Missiles: A New Astrophysical Research Tool," *VIIIth International Astronautical Congress Barcelona 1957*, eds. P. J. Bergeron, Friedrich Hecht (Berlin: Springer, 1958).

66 **"Blown out of proportion"** S. Fred Singer, interview with Allan A. Needell and David DeVorkin, American Institute of Physics, oral-history interviews (April 23, 1991).

66 **After Sputnik was launched** Dickson, *Sputnik*, p. 132.

66 **Magazine features** George Barrett, "Visit with a Prophet of the Space Age," *NYT Magazine* (October, 20, 1957); "The Seer of Space," *Life* (November 18, 1957).

67 **"The rocket that launched"** NBC News special report, "Russians Launch Sputnik 2," 16mm kinescope, NBC Universal Archives [clip no. 511243217] (November 3, 1957).

67 **Democrats with eyes** Senator Henry "Scoop" Jackson, "The Satellite Sky," PBS New York, Robert Stone Productions, Ltd. (December 3, 1990).

67 **Taking von Braun's lead** Robert Caro, *Master of the Senate: The Years of Lyndon Johnson* (New York: Knopf, 2002), p. 1028.

67 **"We have no enemies"** Stephen Ambrose, *Eisenhower: Soldier and President* (New York: Simon & Schuster, 1991), p. 462.

67 "Unless we develop" John D. Morris, "Bigger Rocket an Urgent Need, von Braun and Medaris Warn," *NYT* (December 15, 1957).

68 CBS News's Harry Reasoner Michael D'Antonio, *A Ball, a Dog, and a Monkey: 1957—The Space Race Begins* (New York: Simon & Schuster, 2007), pp. 120–121.

68 In the days that followed "Coverage at Canaveral," *NYT* (December 24, 1957).

69 NBC News used a motorcycle Harold Baker, "Above & Beyond: 'Aw, Hell, Television Is Here,'" *Air & Space* (January 2001).

69 Led by a call "A Jupiter-Based Jubilation," *Life* (February 10, 1958).

70 The mayor joined Alex Thomas, "Wail of Sirens Brings in Era on Space Here," *The Huntsville Times* (February 1, 1958).

71 When he appeared in Washington Ed Buckbee, interview with Robert Stone (April 14, 2015).

71 Von Braun signed Neufeld, *Von Braun*, p. 325.

71 He felt out of place "Top Scientist in Man's Efforts to Shoot Moon," *St. Louis Post-Dispatch* (August 10, 1958).

71 "They should send old" John Elder, "The Experience of Hermann Oberth," *History of Rocketry and Astronautics: Proceedings of the Twenty-Fifth History Symposium of the International Academy of Astronautics* (San Diego: American Astronautical Society, 1997), pp. 302, 308–309.

72 "Our rocketry is good" Milton Esterow, "World Scientists Laud Soviet Shot," *NYT* (January 4, 1959).

72 The Air Force argued Walter McDougall, *The Heavens and the Earth: A Political History of the Space Age* (New York: Basic Books, 1985), p. 166.

72 Its Special Weapons Center William J. Broad, "U.S. Planned Nuclear Blast on the Moon, Physicist Says," *NYT* (May 16, 2000).

74 His announcement in late July "An Explanatory Statement Prepared by the President's Science Advisory Committee," *Introduction to Outer Space* (March 26, 1958).

74 "This is not science fiction" President Dwight Eisenhower, "Statement by the President," *Introduction to Outer Space* (March 26, 1958).

75 Korolev sent a memo Sergei Khrushchev, "How Rockets Learned to Fly," in *Epic Rivalry: The Inside Story of the Soviet and American Space Race* by Von Hardesty and Gene Eisman (Washington: National Geographic, 2007), p. xi.

75 While the Soviet space program's principal Sergei Khrushchev, interview with Robert Stone (February 25, 2015).

75 In fact, during NASA's first year Audra J. Wolf, *Competing with the Soviets: Science, Technology, and the State in Cold War America* (Baltimore: Johns Hopkins, 2103), p. 92.

76 "A propaganda ploy" Sharon Weinberger, *Imagineers of War: The Untold History of DARPA, the Pentagon Agency that Changed the World* (New York: Knopf, 2017), p. 56.

78 When *Real* Dwayne Day, "You've Come a Long Way, Baby!" *The Space Review* (July 15, 2013).

79 **"We have talked about"** "Anecdote of the Week," *Independent Press-Telegram*, Long Beach, CA (August 16, 1964).

79 **"Serious, sober, dedicated, and balanced"** Neufeld, *Von Braun*, p. 338.

79 **By progressing in steps** Buckbee, interview with Robert Stone.

80 **British filmmaker J. Lee Thompson** Steve Chibnall, *J. Lee Thompson* (Manchester: Manchester University Press, 2013), p. 250.

82 **Eisenhower answered without any hesitation** Herbert F. York, *Arms and the Physicist* (Woodbury, NY: American Institute of Physics Press, 1997), p. 147.

CHAPTER THREE: THE NEW FRONTIER

85 **Arthur C. Clarke saw the photographs** Neil McAleer, *Odyssey: The Authorized Biography of Arthur C. Clarke* (London: Gollancz, 1992), p. 168.

86 **"If somebody can"** Hugh Sidey, "Pioneers in Love with the Frontier," *Time* (February 10, 1986).

86 **His science adviser, MIT professor** Mark Erickson, *Into the Unknown Together: The DOD, NASA, and Early Spaceflight* (Montgomery, AL: Air University Press, 2005), p. 205.

87 **John Kennedy once spoke** Newton Minow, interview with the authors (May 12, 2015); *Newton Minow: An American Story*, documentary film by Mike Leonard and Mary Kay Wall (2015).

88 **On his first day** Minow, interview with the authors; *Newton Minow*, Leonard and Wall.

89 **"If we are going"** John W. Finney, "Space Rocket Lag Decried in House," *NYT* (April 18, 1961).

90 **The Pulitzer Prize–winning** Hanson W. Baldwin, "Flaw in Space Policy," *NYT* (April 17, 1961).

91 **On the other hand, the moon** James E. Webb, interview with David DeVorkin, Allan Needell, and Joseph Tatarewicz (April 12, 1985); Glennan-Webb-Seamans Project Interviews 1985–1990, NASM.

91 **On the third and final day** Webb, interview with DeVorkin, Needell, and Tatarewicz.

92 **Leaving the meeting** Webb, interview with DeVorkin, Needell, and Tatarewicz; *The Other Side of the Moon*, BBC Two television documentary (July 20, 1979).

92 **"I want you to know"** Webb, interview with DeVorkin, Needell, and Tatarewicz; *The Other Side of the Moon*.

93 **"If they kill one"** ACC in *Odyssey*, p. 168.

93 **White House press secretary Pierre Salinger** Paul Haney, interview with Sandra Johnson, NASA Johnson Space Center Oral History Project (January 20, 2003).

94 **MOSCOW, April 30 (UPI)** *NYT* (May 1, 1961).

95 **"The Redstone got away"** CBS News (May 5, 1961).

95 **"Come in and watch this!"** John Logsdon, "John F. Kennedy and the Right Stuff," *Quest* 20, no. 2 (2013).

96 **The three television networks** "30 Million Watch Man-in-Space Shot," *Broadcasting* (May 8, 1961): pp 9–10.

98 **While waiting outside the Oval Office** Minow, interview with the authors.

98 **"Number-one television performer"** John F. Kennedy, "Address at the 39th Annual Convention of the National Association of Broadcasters" (May 8, 1961).

99 **"The Gettysburg Address of Broadcasting"** Bob Garfield, *On the Media,* WNYC (June 28, 2017).

100 **Kennedy was exercising a leadership initiative** John Logsdon, interview with the authors (May 11, 2015).

100 **He delivered his speech** "Kennedy Must Convince Public of Value of Moon Shot Project," American Institute of Public Opinion press release, Princeton, NJ, based on a national poll of 1,447 adults (May 31, 1961). Kennedy's approval rating was 77 percent based on polls conducted between May 4–9, 1961.

100 **"I'm not sure"** Robert Gilruth, interview with David DeVorkin and John Mauer (March 2, 1987), Glennan-Webb-Seamans Project Interviews 1985–1990, NASM.

101 **"He was a young man"** Robert Gilruth, interview with Martin Collins and David DeVorkin (October 2, 1986), Glennan-Webb-Seamans Project Interviews 1985–1990, NASM.

101 **"Back in the solar ball park"** Harold M. Schmeck, Jr., "Experts Outline U.S. Space Plans," *NYT* (May 27, 1961).

101 **"When television viewers"** Newton Minow, *Newton Minow,* Leonard and Wall.

102 **"Why don't we"** Edward R. Murrow, letter to James Webb, John F. Kennedy Library microfilm file of USIA director's file (September 21, 1961).

103 **When he returned from the summit** Sergei Khrushchev, interview with Robert Stone (February 25, 2015).

103 **General Dynamics enticed engineers** General Dynamics advertisement, *Aviation Week and Space Technology* (August 14, 1961).

106 **However, he surprised his colleagues** Michael J. Neufeld, "Von Braun and the Lunar-Orbit Rendezvous Decision: Finding a Way to Go to the Moon," *Acta Astronautica.* 63, no. 1–4 (2008).

108 **Days before the launch, Glenn's flight** Bernard Stengren, "Delays in Glenn Flight Proving Costly for Television and Radio," *NYT* (February 17, 1962).

108 **CBS assembled a semi-permanent structure** Alfred Robert Hogan, "Televising the Space Age: A Descriptive Chronology of CBS News Special Coverage of Space Exploration from 1957 to 2003" (master's thesis, University of Maryland, 2005), p. 85.

108 **"If Glenn were only"** Charles Bartlett, interview with Fred Holborn, John F. Kennedy Library, "Charles Bartlett Oral History Interview" (January 6, 1965).

109 **"We have by this time"** NBC News Special Report, "American in Orbit" (February 20, 1962).

109 **Arthur C. Clarke delivered a paper** McAleer, *Odyssey,* p. 172.

110 **"The plain facts of electronic life"** Telstar broadcast (July 23, 1962).

111 **Kennedy looked across the crowd** Minow, interview with the authors.

114 **The university's board of trustees** Richard Paul and Steven Moss, *We Could Not Fail* (Austin: University of Texas Press, 2015), pp. 77–80.

115 **"The first colored man"** Edward R. Murrow memo to the president (April 28, 1962), quoted in A. M. Sperber, *Murrow: His Life and Times* (New York: Freundlich, 1986), p. 657.

116 **Dwight encountered a letter** Ed Dwight, interview with Robert Stone (October 19, 2015).

116 **If he was successful** J. Alfred Phelps, *They Had a Dream: The Story of African-American Astronauts* (Novato, CA: Presidio, 1994), p. 7.

117 **Yeager had assembled his entire staff** Dwight, interview with Robert Stone.

118 **"caught in a buzz saw"** General Chuck Yeager and Leo Janos, *Yeager: An Autobiography* (New York; Bantam Books, 1985), p. 342.

119 **"A twenty-nine-year-old"** NBC News (March 31, 1963), NBC Universal Archive, clip no. 5112432714.

119 **Comedian Dick Gregory** NBC News footage of Los Angeles rally (May 26, 1963), NBC Universal Archives, clip no. 5113378507; Gregory at St. John's Baptist Church in Birmingham, AL (May 20, 1963), recording and transcript at http://americanradioworks.publicradio.org/features/blackspeech/dgregory.html.

120 **"Well, you got an astronaut"** Dwight, interview with Robert Stone.

120 **At the top of the hour** "15 New Astronaut Trainees to Be Selected," news release, NASA Manned Spacecraft Center, MSC-63-095 (June 5, 1963).

121 **Anders was astounded** Bill Anders, interview with the authors (May 13, 2015).

122 **Dr. Charles Lang** Paul and Moss, *We Could Not Fail*, pp. 94–95.

122 **"There are no racial barriers"** Charles L. Sanders, "The Troubles of 'Astronaut' Edward Dwight," *Ebony* (June 1965).

122 **NASA had established two small African** Sunny Tsiao, *"Read You Loud and Clear!" The Story of NASA's Spaceflight Tracking and Data Network* (Washington: NASA, 2008), p. 125.

122 **"All that counts"** "Negro Astronaut Says No Racial Barriers in Space," *Eureka Humboldt Standard* (April 8, 1963).

123 **In an attempt to mollify** Dwight, interview with Robert Stone.

123 **The accounts never mentioned** John Charles, "A Hidden Figure in Plain Sight," *The Space Review* (June 12, 2017).

123 **As part of their training** Andrew Chaikin, *A Man on the Moon: The Voyages of the Apollo Astronauts* (New York: Viking, 1994), pp. 611–612.

124 **He was confident** Dwight, interview with Robert Stone.

124 **"Without qualification"** Phelps, *They Had a Dream*, p.21; UPI wire story (June 3, 1965).

124 **"Why are these people"** Anders, interview with the authors.

125 **"Was there a Negro boy"** Audio recording of NASA press conference (October 18, 1963), NASM, Michael Kapp Collection.

125 **Reinforced-concrete family bomb shelter** Michael J. Neufeld, *Von Braun: Dreamer of Space, Engineer of War* (New York: Knopf, 2007), p. 382.

125 Newton Minow hastily departed *Newton Minow*, Leonard and Wall.

126 In Minow's estimation Minow, interview with the authors.

127 Before addressing the General Assembly Dodd L. Harvey and Linda C. Ciccoritti, *U.S.-Soviet Cooperation in Space* (Washington: Center for Advanced International Studies, University of Miami, 1974), p. 122.

127 *Life* magazine was forced "How We Can Join In a Moon Trip," *Life* (October 11, 1963).

127 Arizona Senator Barry Goldwater "Goldwater Can't See Russ Trust," *San Mateo Times* (September 21, 1963).

127 Privately, Nikita Khrushchev discussed Sergei Khrushchev, interview with Robert Stone.

128 Accounts of his mood Logsdon, "John F Kennedy and the Right Stuff," *Quest*.

128 "Last Saturday at Cape Canaveral" John F. Kennedy, remarks at the dedication of the Aerospace Medical Health Center, San Antonio, TX (November 21, 1963).

129 He caught Florida senator Spessard Holland Lyndon Johnson telephone call to Spessard Holland (November 27, 1963), audio recording, Lyndon Johnson Library, tape K6311.03, PNO 32.

129 "We're getting ready" Lyndon Johnson telephone call to C. Farris Bryant (November 27, 1963), audio recording, Lyndon Johnson Library, tape K6311.03, PNO 34.

131 Ensuring that no NASA money Public Law 88-215, 88th Congress, 1st session (December 19,1963).

CHAPTER FOUR: WELCOME TO THE SPACE AGE

133 In his brief speech "Talk of the Town," *The New Yorker* (September 26, 1964), p. 38; James Webb, "Remarks on the Dedication of the Hall of Science" (September 9, 1964), http://www.nywf64.com/spacpark10.shtml.

134 He was an asute observer W. Henry Lambright, *Powering Apollo: James E. Webb of NASA* (Baltimore: Johns Hopkins, 1995), p. 27.

136 "Do something!" "Shouts Mar Johnson's Talk at Pavilion," *NYT* (April 23, 1964).

138 "A brilliant woman" James Webb, interview with Martin Collins (October 15, 1985), Glennan-Webb-Seamans Project Interviews 1985–1990, NASM.

138 "Planned disequilibrium" W. Henry Lambright, "Leading NASA in Space Exploration: James E. Webb, Apollo, and Today," in *Leadership and Discovery*, eds. George R. Goethals and J. Thomas Wren (New York: Palgrave Macmillan, 2009).

138 Von Braun communicated Roger Launius, "Comments on a Very Effective Communications System: Marshall Space Flight Center's Monday Notes," *Roger Launius's Blog* (February 28, 2011), https://launiusr.wordpress.com/2011/02/28/comments-on-a-very-effective-communications-system-marshall-space-flight-center%E2%80%99s-monday-notes/.

139 "How [can] anything" James E. Webb, interview with T. H. Baker (April 29, 1969), Lyndon Baines Johnson Library Oral History Collection.

139 **"Our people [have] to"** James E. Webb, interview with Martin Collins and Allan Needell (November 4, 1985), Glennan-Webb-Seamans Project Interviews 1985–1990, NASM.

140 **"Webb's got all the money"** Lambright, *Powering Apollo*, pp. 112–113.

140 **"We're not getting any fancy jets"** Robert C. Seamans, Jr., *Aiming at Targets* (Washington: NASA History Office, 1996), p. 101.

142 **Recovery aircraft had picked up** Loyd S. Swenson, Jr., James M. Grimwood, and Charles C. Alexander, *This New Ocean: A History of Project Mercury* (Washington: NASA History Office, 1998), pp. 446–460.

142 **"I accept your offer"** Billy Watkins, *Apollo Moon Missions: The Unsung Heroes* (Westport, CT: Praeger, 2006), p. 55.

143 **'I'll tell you one goddamned thing'** Brian Duff, interview with John Mauer (April 24, 1989), Glennan-Webb-Seamans Project Interviews 1985–1990, NASM.

143 **"The best PR man"** Duff, interview with Mauer (May 1, 1989), Glennan-Webb-Seamans Project Interviews 1985–1990, NASM.

143 **In one such instance** Jim Hicks, "Hardest Rendezvous of All—On the Ground," *Life* (October 1, 1965): pp. 113–116.

144 **"We're not the Soviets"** Watkins, *Apollo Moon Missions*, p. 52.

144 **During one internal presentation** Duff, interview with Mauer (April 24, 1989), Glennan-Webb-Seamans Project Interviews 1985–1990, NASM.

145 **"It would be better"** Gerard J. Degroot, *Dark Side of the Moon: The Magnificent Madness of the American Lunar Quest* (New York: NYU Press, 2006), pp. 183–184. *Marooned* wasn't produced for another five years, by which time the story had been revised to follow the plight of a later Apollo Applications mission, in the early 1970s. When the film finally went before the cameras, NASA did provide some technical assistance.

146 **He once joked that Lange's art** David Larson, "Harry Lange," *Guardian* (July 7, 2008).

147 **Kubrick sent a note** Tom Turner, letter to Stanley Kubrick with handwritten reply from Kubrick (January 17, 1965), in David Meerman Scott and Richard Jurek, *Marketing the Moon* (Cambridge: MIT Press, 2014), p. 14.

148 **Opinion polls at the time** Roger D. Launius, "Public Opinion Polls and Perceptions of U.S. Human Spaceflight," *Space Policy* 19, no 3 (August 2003): pp. 163–175.

148 **The former astronaut observed** Robert Dallek, "Johnson, Project Apollo, and the Politics of Space Program Planning," in *Spaceflight and the Myth of Presidential Leadership*, eds.Roger D. Launius and Howard E. McCurdy (Urbana, IL: University of Illinois Press, 1997), p. 75.

149 **A leading Democratic House member** "Boggs Says Goldwater Would Peril Space Race," *NYT* (August 2, 1964).

149 **"The shackles of Earth"** Lyndon Johnson, "The New World of Space," speech delivered in Seattle, WA (May 10, 1962), *Proceedings of the Second National Conference on the Peaceful Uses of Space* (Washington: NASA, 1962), pp. 29–30.

153 **Novelist James Salter** James Salter, *Burning the Days: Recollection* (New York: Random House, 1997), pp. 59, 268.

155 "The Christopher Columbuses of the twentieth century" Lady Bird Johnson, *A White House Diary* (Austin: University of Texas, 2007), p. 288.

155 "Men who have worked together" Lady Bird Johnson, *Diary*, p. 298.

155 On the front cover *A Guide to Careers in Aero-Space Technology* (Washington: NASA, 1966).

155 They attended the dedication "Astronauts Among 1,000 Who See Dedication of U-M Space Center," *Ann Arbor News* (June 15, 1965).

156 Johnson complained to Webb Lambright, "Leading NASA in Space Exploration," in *Leadership and Discovery*, eds. Goethals and Wren, p. 86.

157 Webb found himself Duff, interview with Mauer (April 24, 1989), Glennan-Webb-Seamans Project Interviews 1985–1990, NASM; Richard Paul and Steven Moss, *We Could Not Fail* (Austin: University of Texas Press, 2015), pp. 203–205.

158 A ruthless and dangerous demagogue Paul and Moss, *We Could Not Fail*, p. 214.

158 In his speech, von Braun Ben A. Franklin, "Wallace Is Given a NASA Warning; He Is Told on 'Truth' Tour to Liberalize His Policies," *NYT* (June 9, 1965).

159 In a profile Ben A. Franklin, "Von Braun Fights Alabama Racism; Scientist Warns State U.S. Might Close Space Center," *NYT* (June 14, 1965).

160 Dwight responded with a public statement Gladwin Hill, "Negro Pilot Finds Bias in Air Force; But Absolves NASA on Being Dropped as Astronaut," *NYT* (June 3, 1965).

160 "I have no idea" Ed Dwight, interview/press conference with ABC News (June 3, 1965), 16mm news film, number A14059.

161 "Now I understand" Ed Dwight, interview with Robert Stone (October 19, 2015).

162 "U.S.-Soviet box score" "U.S.-Soviet Box Score on Astronauts' Flights," *NYT* (July 21, 1966).

162 The world learned his name "The Secret Scientist," *NYT* (January 20, 1966).

167 He couldn't forget Barton C. Hacker and James M. Grimwood, *On the Shoulders of Titans: A History of Project Gemini* (Washington: NASA, 1977), p. 318.

168 "One of the year's best" Allen Rich, "TV Week," *Independent Star-News*, Pasadena, CA (March 27, 1966).

168 Network switchboards registered "TV Fans Protest Canceling of Shows for Gemini Report," *NYT* (March 17, 1966).

168 "Too long, boring" Ernie Kreilino, "A Closer Look," *Van Nuys Valley News* (May 3, 1966), p. 35.

168 Armstrong appeared depressed James R. Hansen, *First Man: The Life of Neil A. Armstrong* (New York: Simon & Schuster, 2005), p. 265.

169 One of his earliest memories Barack Obama, "America Will Take the Giant Leap to Mars," CNN.com (October 11, 2016), https://www.cnn.com/2016/10/11/opinions/america-will-take-giant-leap-to-mars-barack-obama/index.html.

170 Some reporters were convinced Mark Bloom, interview with Robert Stone (March 25, 2015).

171 "One of the largest sacred cows" Dale L. Cressman, "Fighting for Access: ABC's 1965–66 Feud with NASA," *American Journalism* 24, no. 3 (2007).

171 **Jules Bergman believed** "Are Newsmen Hampered?" *Broadcasting* (April 18, 1966): p. 56.

172 **Something of a rude awakening** Lambright, *Powering Apollo*, p. 140; Donald Janson, "Webb Backs Cost of Space Program," *NYT* (December 6, 1966).

173 **What irked Lehrer** Andrew Robinson, "Tom Lehrer at 90: A Life of Scientific Satire," *Nature* 556, no. 7699 (April 5, 2018): pp. 27–28.

173 **Arthur C. Clarke provides** ACC, *Astounding Days* (London: Gollancz, 1989), p. 162.

174 **In private they joked** Bloom, interview with Robert Stone.

174 **In a public interview** Evert Clark, "NASA Chief Urges Space Planning Now for Post-Moon Era," *NYT* (May 30, 1966).

175 **Rented a large pool** Tom Jones, "Where NASA Learned to Spacewalk," *Air & Space* (September 2014).

175 **"Victory over space"** "Again a Victory over Space," *NYT* (November 16, 1966).

176 **The excitement surrounding** "Again a Victory," *NYT*.

177 **"The months ahead"** "Johnson Cautions on Moon Mission: Apollo's Task 'Complicated' He Honors Astronauts," *NYT* (November 24, 1966).

CHAPTER FIVE: EARTHRISE

180 **Once each month** George Alexander, interview with Robert Stone (June 2, 2015).

181 **The small El Lago suburb** Bill Anders, interview with the authors (May 13, 2015).

182 **In the midst of** Robert Seamans, interview with Martin Collins and Henry Lambright (December 16, 1988), Glennan-Webb-Seamans Project Interviews 1985–1990, NASM.

183 **NBC's Hackes explained** NBC News Special Report (January 28, 1967), black-and-white kinescope, NBC Universal Archives, clip no. 51A17171_S01.

183 **"This is a time"** CBS News Special Report (January 27, 1967), https://www.c-span.org/video/?422730-1/cbs-news-special-report-apollo-1-disaster.

185 **The astronaut families gathered** James R. Hansen, *First Man: The Life of Neil A. Armstrong* (New York: Simon & Schuster, 2005), p. 307.

185 **Anonymous air-traffic controllers** Frank Borman and Robert J. Serling, *Countdown: An Autobiography* (New York: Silver Arrow/William Morrow, 1988), p. 171.

186 **A strong odor of burned paper** Alexander, interview with Robert Stone.

187 **"I want you to do it"** James E. Webb, interview with T. H. Baker (April 29, 1969), Lyndon Baines Johnson Library Oral History Collection.

188 **When reporters questioned Julian Scheer** Mark Bloom, interview with Robert Stone (March 25, 2015).

190 **Deemed the memo not worth pursuing** W. Henry Lambright, *Powering Apollo: James E. Webb of NASA* (Baltimore: Johns Hopkins University Press, 1995), p. 158.

190 **"We are the apostles"** James Webb, memo to Robert Seamans (April 1, 1967), in *Powering Apollo*, p. 159.

192 **"I don't want you doing anything"** Frank Borman, interview with Catherine Harwood (April 13, 1999), NASA Johnson Space Center Oral History Program, https://www.c-span.org/video/?293191-1/frank-borman-oral-history-interview.

192 **"You are asking us"** William J. Normyle, "NASA Revises Manned Flight Plan," *Aviation Week and Space Technology* (April 24, 1967), p. 29; *Investigation into Apollo 204 Accident: Hearings Before the Subcommittee on NASA Oversight of the Committee on Science and Astronautics, U.S. House of Representatives, Ninetieth Congress* (Washington: Government Printing Office, 1967), p. 446.

193 **The members were prepared to vote** Lambright, *Powering Apollo*, p, 169.

193 **A West Virginia representative** "Calls for NASA Shake-Up," *Virgin Islands Daily News* (April 11, 1967); *Investigation into Apollo 204 Accident,* p. 457.

194 **"Bunch of bums"** Thomas O. Paine, interview with Robert Sherrod (October 7, 1971), in *"Before This Decade Is Out . . .": Personal Reflections on the Apollo Program,* ed. Glen E. Swanson (Washington: NASA, 1999), p. 36.

194 **By mid-1967** Kathleen Weldon, "Fly Me to the Moon: The Public and NASA," *Huffington Post* (February 25, 2015).

194 **"Do you wish"** John T. Dugan (as John Kingsbridge), with Gene L. Coon, Gene Roddenberry, and John Meredyth Lucas, "Return to Tomorrow," *Star Trek*, NBC television (February 8, 1968).

198 **But half a century later** Dwayne Day, "From the Shadows to the Stars: James Webb's Use of Intelligence Data in the Race to the Moon," *Air Power History* 51, no. 4 (Winter 2004).

202 **Robert Gilruth argued** Robert Gilruth, interview with James Burke in *The Other Side of the Moon,* BBC Two television documentary (July 20, 1979).

202 **Chose to disobey Webb's request** Thomas O. Paine, interview with Burke in *The Other Side of the Moon.*

203 **Webb could carry on the fight** James E. Webb, interview with Burke in *The Other Side of the Moon.*

204 **The animals were returned** "Zond 5," *NASA Space Science Data Coordinated Archive* (March 2017), https://nssdc.gsfc.nasa.gov/nmc/spacecraftDisplay.do?id=1968-076A.

204 **"Three Apollo astronauts may"** "Apollo Plans Set for Circling Moon," *NYT* (September 15, 1968).

205 **"Program in decline"** "Lost: One Space Booster," *Newsweek* (September 23, 1968).

206 **A KGB colonel** Robert Pear, "Double Agent, Revealed by FBI, Tells of Technique," *NYT* (March 4, 1980); Arlin Crotts, *The New Moon: Water, Exploration, and Future Habitation* (New York: Cambridge University Press, 2014), p. 56, n. 30.

207 **Anders dutifully** Anders, interview with the authors.

207 **He would rather die than screw up** Anders, interview with the authors; Anders, interview with Jim Hartz, in "LBJ Library & Museum Presents Apollo 8 Reunion with Frank Borman, James Lovell, William Anders" (April 23, 2009), https://www.youtube.com/watch?v=Wa5xoT-peeo.

208 **Jules Bergman pressed Borman** Frank Borman, comments at NASA press con-

ference (December 7, 1968), KHOU Channel 11 Collection, Box 6804, Reel 23, Houston Area Digital Archives, Houston Public Library.

209 **The political protests seemed** Frank Borman, interview with the authors (June 11, 2015).

210 **"It's unfortunate"** Walter Cunningham, *The All-American Boys* (New York: Macmillan, 1977), pp. 62–63.

210 **Manhattan Project physicist Ralph Lapp** "Physicist Asks Delay of Apollo 8," UPI, *Detroit Free Press* (December 9, 1968); Ralph Lapp, interview in "The Coming Trip Around the Moon," *The New Republic* (December 14, 1968); "Apollo 8 a Death Trap? Borman Denies It," AP, *Tucson Daily Citizen* (December 10, 1968).

210 **Option of taking suicide pills** "Historic Apollo 8 Flight Perfect in Early Stages," *Calgary Herald* (December 21, 1968).

211 **Bill Anders sensed the rocket moving** Anders, interview with the authors; Anders in "An Evening with the Apollo 8 Astronauts: Annual John H. Glenn Lecture Series" (November 13, 2008), NASM, https://www.youtube.com/watch?v=Q2h_FtLzrrU.

212 **"That's the big decision!"** CBS News Special Report, "The Flight of Apollo 8" (December 21, 1968).

212 **"Jeez, there's got to be"** Michael Collins, interview with Michelle Kelly (October 8, 1997), NASA Johnson Space Center Oral History Project.

213 **"The Death Watch"** Susan Borman, interview in "Astronaut Wives Club," BBC Radio 4 (November 9, 2007).

213 **Late that evening** Andrew Chaikin, *A Man on the Moon* (New York: Penguin, 1994), pp. 124–125; Borman and Serling, *Countdown*, p. 295; Robert Zimmerman, *Genesis: The Story of Apollo 8, the First Manned Flight to Another World* (New York: Four Walls, Eight Windows, 1998), pp. 177–178.

214 **Frances "Poppy" Northcutt** Poppy Northcutt, interview with the authors (June 10, 2015).

219 **The Moon's landscape monotonous** Anders, interview with the authors.

219 **It had never been discussed** Anders, interview with the authors; *Apollo 8 Onboard Voice Transcription* (Houston: Manned Spacecraft Center, 1969).

220 **Margaret Mead declared** Margaret Mead, interview with Wilton S. Dillon (September 13, 1974), Smithsonian Institution, Washington, DC; Mead made a similar statement in the film *Our Open-Ended Future* (Moffett Field, CA: NASA Ames Research Center, 1973) as part of a NASA lecture series "The Next Billion Years: Our Future in Cosmic Perspective."

221 **"Peace on Earth"** Billy Watkins, *Apollo Moon Missions: The Unsung Heroes* (Westport, CT: Praeger, 2006), p 70.

221 **"We've got to do it up"** *Apollo 8 Onboard Voice Transcription*, p. 173.

222 **Cronkite momentarily thought** Walter Cronkite, interview with Kevin Michael Kertscher, "The Making of 'Race to the Moon': Apollo 8 Documentary Producer Tells All," Space.com (October 20, 2005); "Telecasts from Apollo 8," http://www.pbs.org/wgbh/americanexperience/features/moon-telecasts-apollo-8/.

223 **"No. Leave it off. Great."** *Apollo 8 Onboard Voice Transcription*, p. 196.

223 **"Well . . . quite a finish"** CBS News Special Report, "The Flight of Apollo 8" (December 24, 1968).

224 **Novelist William Styron was celebrating** William Styron, introduction in *The View from Space: American Astronaut Photography 1962–1972* by Ron Schick and Julia Van Haaften (New York: Clarkson Potter, 1987).

224 **Scheer took a call** Robert J. Donovan and Raymond L. Scherer, *Unsilent Revolution: Television News and American Public Life, 1948–1991* (New York: Cambridge University Press, 1992), p. 51; Paul G. Dembling, interview with Edward S. Goldstein, Gregory C. La Rosa, and David S. Schuman in "Present at the Creation: Paul G. Dembling, Author of NASA's Founding Legislation," *NASA 50th Magazine: 50 Years of Exploration and Discovery* (2008).

226 **Associated Press released** "Space Race," AP, *The Post-Crescent,* Appleton, WI (December 27, 1968); Tom Wicker, "In The Nation: Walter Mitty and Mount Everest," *NYT* (December 29, 1968).

226 **Time's editors** Zimmerman, *Genesis,* p. 235.

227 **Northcutt started to receive fan mail** Northcutt, interview with the authors.

229 **Featured her in a cover story** "Women Arise," *Life* (September 4, 1970): p. 20.

CHAPTER SIX: MAGNIFICENT DESOLATION

231 **"We saw the Earth"** "3 Moon Voyagers Are Hailed Here in Huge Turnout," *NYT* (January 11, 1969).

234 **Struggled with depression** Buzz Aldrin, interview with Robert Stone (November 19, 2014).

235 **There was a standoff** Aldrin, interview with Robert Stone.

237 **As a compromise** Wilson P. Dizard, Jr., *Inventing Public Diplomacy: The Story of the U.S. Information Agency* (Boulder, CO: Lynne Rienner, 2004), p. 111; Anne M. Platoff, "Flags in Space: NASA Symbols and Flags in the U.S. Manned Space Program," *The Flag Bulletin* 46, no. 5–6 (September–December 2007); Billy Watkins, *Apollo Moon Missions: The Unsung Heroes,* (Westport, CT: Praeger, 2006), p. 57.

238 **The journals of Lewis and Clark** George Plimpton, "Neil Armstrong's Famous First Words," *Esquire* (December 1983): p. 118; Julian Scheer, letter to George Low (March 12, 1969).

239 **"We should dream no small dreams"** "3 Moon Voyagers Are Hailed Here in Huge Turnout," *NYT.*

240 **Paine had been at Cape Kennedy** Thomas O. Paine, interview with Eugene M. Emme (September 3, 1970), NASA Oral History, Folder 4186.

240 **The comments had been made** Robert Reinhold, "Kennedy Puts Earth Needs Ahead of Space Program," *NYT* (May 20, 1969); "NASA Chief Hits Talk by Kennedy," *The Washington Post* (May 21, 1969), p. A12.

240 **Muskie believed** "Kennedy Has Inside Track for 1972, Muskie Thinks," *Nashua Telegraph* (July 1, 1969), p. 29.

241 **Suggested a symbolic gesture** Paine, interview with Emme.

245 **It crashed to the ground** "Soviets Suffer Setbacks in Space," *Aviation Week and Space Technology* (November 17, 1969), pp. 26–27.

246 **"Unjust, evil, and futile"** Martin Luther King, Jr., speech, "Why I Am Opposed to the War in Vietnam" (April 30, 1967).

246 **"God's children"** Martin Luther King, Jr., speech, "Where Do We Go from Here?" (August 16, 1967).

247 **Tom Paine was attending** Thomas O. Paine, "We Are Also Americans," *21st Century* (May–June 1989), pp. 30–31; Julian Scheer, "The Sunday of the Space Age," *The Washington Post* (December 8, 1972), p. A26.

247 **Position paper** "The Space Program," Southern Christian Leadership Conference Position Paper (July 14, 1969), SCLC Records 1964–2003, Manuscript, Archives, and Rare Book Library, Emory University.

250 **Paine said he had come** CBS News, 16mm film footage of Poor People's Campaign protest (July 15, 1969).

251 **"Watching Columbus"** John Logsdon, interview with authors (May 11, 2015).

252 **He boldly predicted** "Lovell Says Soviet Attempts to Extract Specimens of Moon," *NYT* (July 16, 1969).

253 **"I suddenly understood"** Theo Kamecke, interview with Robert Stone (May 5, 2015).

253 **In newsrooms** Walter Cronkite, CBS News, "Man on the Moon: The Epic Journey of Apollo 11" (July, 16, 1969).

254 **Its struggle to escape Earth's gravity** Lyndon Johnson, interview with Cronkite, "Man on the Moon."

255 **"I really forgot"** Al Rossiter, Jr., "Apollo 11 Blasts Off for Historic Moon Voyage," UPI wire story (July 16, 1969).

255 **"Phooey"** "Set Earth Priorities," *Florida Today from Cocoa, Florida* (July 17, 1969), p. 11A.

256 **Sargent Shriver** "JFK Dream Comes True," *Florida Today from Cocoa, Florida* (July 17, 1969), p. 11A; Richard S. Lewis, *Appointment on the Moon* (New York: Viking, 1969), p. 504.

258 **"Do you think"** ACC, interview with Cronkite, "Man on the Moon."

261 **"The hopes and the dreams"** Gene Kranz, *Failure Is Not an Option: Mission Control from Mercury to Apollo 13 and Beyond* (New York: Simon & Schuster, 2000), p. 283.

263 **Armstrong later said** Neil Armstrong, quoted in *Apollo 11: Technical Crew Debriefing, July 31, 1969* (Houston: Manned Spacecraft Center, 1969), p. 60.

264 **"A symbolic act"** Lewis Mumford, "No: 'A Symbolic Act of War . . . '" *NYT* (July 21, 1969), p. 6.

264 **"How can anybody"** CBS News, "Man on the Moon," (July 20, 1969).

264 **"Old-fashioned humanist"** CBS News, "Man on the Moon," (July 20, 1969).

265 **Astronauts deemed** Neil Armstrong, "The Moon Had Been Awaiting Us a Long Time," *Life* (August 22, 1969): p. 25. Buzz Aldin, interview with Eric M. Jones (1991), https://www.hq.nasa.gov/alsj/a11/a11.evaprep.html.

265 **He grinned** John Noble Wilford, "Apollo Crew Appears Calm 11 Days Before the Mission," *NYT* (July 6, 1969).

265 **During a lull** Dean Armstrong, interviewed in *Neil Armstrong—First Man on the Moon,* BBC Two (December 30, 2012).

265 **Truman Capote** William H. Honan, "Le Mot Juste for the Moon," *Esquire* (July 1969).

267 **"I need you"** George Alexander, interview with Robert Stone (June 2, 2015).

268 **"You're not"** George Plimpton, "Neil Armstrong's Famous First Words," *Esquire* (December 1983): pp. 116–118.

268 **Lunar scientists theorized** Leonard Reiffel, interview with authors (May 12, 2015).

270 **"I didn't hear that"** Julian Scheer, "What About God? NASA Ignored Nixon's Order," *Orlando Sentinel* (July 20, 1989).

271 **"Neil made me a promise"** Kamecke, interview with Robert Stone. Kamecke said, "I never revealed it because I just thought it would be crude. I could imagine what the media would do with such a thing, since the media's attitude was very weird in those days. A lot of them didn't think too much about the space program."

271 **One-fifth of the world's population** Paul Harris, "Man on the Moon: Moment of Greatness that Defined the American Century," *Guardian* (August 25, 2012).

272 **Forcefully advised** Frank Borman, "Memorandum to the President" (July 14, 1969), Richard Nixon Presidential Library and Museum; Frank Borman, interview with the authors (June 11, 2015).

273 **"I believe firmly"** Ray Bradbury, interview with Mike Wallace, CBS News, "Man on the Moon" (July 21, 1969).

274 **Joyful and in tears** Ray Bradbury, speech delivered upon being awarded the National Book Foundation's Medal for Distinguished Contribution to American Letters (November 15, 2000); see also Bradbury's public comments during appearance at Comic-Con (July 29, 2009).

274 **"The tune of lunar gravity"** Vladimir Nabokov, interview with James Mossman, Review, BBC Two (October 4, 1969).

274 **Pointedly observed** Harry Schwartz, "Capitalist Moon or Socialist Moon?" *NYT* (July 21, 1969).

274 **"If the government"** Ayn Rand, "Apollo 11," *Ayn Rand Reader* (New York: Penguin, 1999), p. 133.

275 **"It means nothing"** "Reactions to Man's Landing on the Moon Show Broad Variations in Opinions," *NYT* (July 21, 1969).

275 **"flying cover"** Paine, interview with Emme.

275 **CBS's Bill McLaughlin** *10:56:20 PM EDT 7/20/69* (New York: Columbia Broadcasting System, 1970), pp. 88, 111.

276 MAROONED! Mark Bloom, interview with Robert Stone (March 25, 2015).

278 **An editorial writer** Bob Evans, "Apollo 11 Crew Ready for Moon Landing," segment for Sunday Morning, Canadian Broadcasting Company radio broadcast (July 20, 1969), archived: http://www.cbc.ca/archives/entry/apollo-11-crew-ready-for-moon-landing.

279 "They are the sons" Hugh Sidey, "Marshalling the Good Guys," *Life* (August 21, 1970): p. 2B.

280 Cape Town *Sunday Times* Carin Bevan, "Putting Up Screens: A History of Television in South Africa 1929–1976," (MHCS diss., University of Pretoria, 2008), pp. 111–119.

280 Pro-Soviet news outlets "The Moon Landing through Soviet Eyes: A Q&A with Sergei Khrushchev, son of former premier Nikita Khrushchev," *Scientific American* (July 16, 2009); *Astronautics and Aeronautics, 1969: Chronology on Science, Technology and Policy* (Washington: NASA, 1970), SP-4014, p. 241.

281 East German television Sven Grampp, "Watching Television, Picturing Outer Space and Observing the Observer Beyond: The First Manned Moon Landing as Seen on East and West German Television," in *Television Beyond and Across the Iron Curtain* by Kirsten Bönker, Julia Obertreis, and Sven Grampp (Newcastle on Tyne: Cambridge Scholars Publishing, 2016), pp. 80–86.

281 Concerned about the tremendous expense Asif A. Siddiqi, *Challenge to Apollo: The Soviet Union and the Space Race, 1945–1974* (Washington: NASA, 2000), p. 426.

282 Bill Plante CBS News, *10:56:20 PM EDT 7/20/69* (New York: Columbia Broadcasting System, 1970), pp. 144–145; Art Peters, "75,000 Miss Moon Landing; Rock in Rain to Motown 'Soul' Music," *Philadelphia Tribune* (July 26, 1969), p. 22.

282 "The first non-racist moment" *Chicago Defender* (July 21, 1969).

282 May have seen Nona Smith's letter Nona E. Smith, "Pride in Identifying," *NYT* (August 9, 1969).

283 A reporter present noticed "L.A. State Dinner Enjoyed by Black Figures," *Jet* (August 28, 1969), p. 8.

283 Lawrence's introduction Gladwin Hill, "Negro Among Four Chosen as Crew of Manned Orbiting Laboratory," *NYT* (July 1, 1967).

284 "The overview effect" Frank White, *The Overview Effect: Space Exploration and Human Evolution* (Boston: Houghton Mifflin, 1987).

284 Poppy Northcutt, who served Northcutt, interview with the authors.

284 A new era of exploration Julian Scheer, interview in *One Small Step: Man on the Moon,* BBC Two television documentary (June 28 and June 30, 1994); Julian Scheer, "The Sunday of the Space Age," *The Washington Post* (December 8, 1972), p. A26.

CHAPTER SEVEN: THE FINAL FRONTIER

287 "Forward-thrusting space program" Spiro Agnew, interview with Roger Mudd, CBS News, "Man on the Moon: The Epic Journey of Apollo 11" (July 20, 1969).

287 Had given her nightmares Marilyn Lovell, interview and DVD feature commentary, *Apollo 13* (Universal City, CA: Universal, 2005).

287 Began worrying John Logsdon, interview with the authors (May 11, 2015).

289 **An opinion poll indicated** "U.S. Spending: New Priorities," *Newsweek* (October 6, 1969), p. 46; *Congressional Record* (October 7, 1969), p. 28837.

290 **"I don't think that the moon"** Mike Geoghegan, letter to Joe Consolino, "Memo to General Manager of the Trade Division" (July 20, 1970), Little, Brown & Company papers, Houghton Library, Harvard University.

290 **"Could we advertise"** Norman Mailer, letter to Ned Bradford (August 10, 1970), Little, Brown & Company papers, Houghton Library, Harvard University.

290 **"Too many people"** Wernher von Braun, interview with John Logsdon (August 25, 1970), in *The Space Shuttle Decision* by T. A. Heppenheimer (Washington: NASA, 1999), p. 169.

291 **Global supercomputer and telecommunications network** The first message transmitted between two computers on ARPANET, the forerunner of the Internet, occurred on October 29, 1969, eight months before the Wallops Island conference. By the time of the conference in June 1970, nine computer systems were connected to ARPANET.

291 **"Like buying a Rolls-Royce"** Richard D. Lyons, "New Cuts for Apollo: 'No Gas for the Rolls-Royce?'" *NYT* (September 6, 1970), p. 107.

292 **Astronaut Tom Stafford** "NASA Cutbacks," NBC Evening News (September 2, 1970).

295 **Clarke confided to Bernstein** Jeremy Bernstein, "The Grasshopper and His Space Odyssey: A Scientist Remembers the Celebrated Science Fiction Writer Arthur C. Clarke," *American Scholar* (Summer 2008).

296 **Semaphore and smoke-signal era** ACC, interview filmed at the Massachusetts Institute of Technology Conference on the Centennial of the Invention of the Telephone (March 9, 1976), ATT 16mm film: https://www.youtube.com/watch?v=D1vQ_cBof4w.

296 **Cranks were littering** ACC, interview with Malcolm Kirk, *Omni* (March 1979).

296 **Motivated by anger** Bill Kaysing, interview with Nardwuar, aka John Ruskin (February 16, 1996), https://nardwuar.com/vs/bill_kaysing/.

296 **As early as 1968** *Thirty-Minute Theatre,* "The News Benders," BBC Two (January 10, 1968).

296 **"In another couple of years"** Norman Mailer, interview with Studs Terkel (January 29, 1971).

297 **"Mass hoodwinking"** Norman Mailer, "A Fire on the Moon," *Life* (August 29, 1969).

298 **One went so far** Bill Anders, interview with the authors (May 13, 2015).

301 **Marvelous achievement** Freeman Dyson, interview with Robert Stone (March 19, 2015).

APPENDIX: AFTER A FEW MORE REVOLUTIONS AROUND THE SUN

305 **Buzz Aldrin found it** Edwin E. "Buzz" Aldrin, Jr., and Wayne Warga, *Return to Earth* (New York: Random House, 1973), pp. 38, 43, 45.

305 **He was surprised** Bill Anders, interview with the authors (May 13, 2015).

306 **"Here we are"** Ron Judd, "With a View from Beyond the Moon, an Astronaut Talks Religion, Politics and Possibilities," *Seattle Times Pacific NW Magazine* (December 7, 2012).

306 **Informed his other colleagues** George Alexander, interview with Robert Stone (June 2, 2015).

307 **Moral substitute for war** Lou Cannon, "The Puzzling Politics Of Jerry Brown," *The Washington Post* (February 5, 1978).

307 **"I am too busy proving"** ACC, correspondence to James Randi (July 11, 2001; published July 20, 2001), http://archive.randi.org/site/jr/07-20-01.html.

307 **Experience was not pleasant** Mordecai Lee, "The Astronaut and Foggy Bottom PR: Assistant Secretary of State for Public Affairs Michael Collins, 1969–1971," *Public Relations Review* 33, no. 2 (2007).

308 **"I heard we've got lots"** Clyde Haberman, "Dick Gregory, 84, Dies; Found Humor in the Civil Rights Struggle," *NYT* (August 19, 2017).

308 **Among the first prominent celebrities** "Group: Space Program Moon Shots a Perpetual Star Trek," *Longview Texas News-Journal* (July 21, 1994).

308 **He used a FOIA request** Lynn Darling, "A Rebel Redeemed," *The Washington Post* (April 23, 1980).

308 **NASA received a letter** Courtney G. Brooks, James M. Grimwood, and Loyd S. Swenson, *Chariots for Apollo: The NASA History of Manned Lunar Spacecraft to 1969* (Washington: NASA, 1979), p. 331.

309 **"I resent it"** John Carmody, "Astronaut Pans Apollo 13 Movie," *Victoria Advocate* (March 2, 1974); James Lovell, letter to James Fletcher (February 11, 1974).

310 **He remained interested** John Elder, "The Experience of Hermann Oberth," in *History of Rocketry and Astronautics: Proceedings of the Twenty-Fifth History Symposium of the International Academy of Astronautics* (San Diego: American Astronautical Society: San Diego, 1997); "Wahre Liebe," *Der Spiegel* (September 8, 1965).

310 **"Space truck"** John M. Broder, "NASA Is Viewed as Deeply Troubled, Uncertain of Its Goals and Purpose," *Los Angeles Times* (October 2, 1988).

311 **Powers argued for diverting money** "Shorty Powers Preaches Peace," *Toledo Blade* (May 12, 1970).

311 **A reported hundred-thousand-dollar contract** Edwin Diamond, "The Dark Side of the Moonshot Coverage," *Columbia Journalism Review* 8, no. 3 (Fall 1969).

311 **"Walter to Walter coverage"** CBS advertisement, *NYT* (July 24, 1969).

312 **In poor taste** Norman Mailer, interviewed on *The Dick Cavett Show* (July 21, 1971).

312 **"Global warming, if it occurs"** S. Fred Singer, "Global Apocalypse Fantasy," *The Washington Times* (November 26, 1997).

312 **"Risk for lung cancer"** S. Fred Singer, "Anthology of 1995's Environmental Myths," *The Washington Times* (February 11, 1996).

312 **"Granddaddy of fake 'science'"** Tim Dickinson, "The Climate Killers," *Rolling Stone* (January 2010).

312 **Turned down positions** W. Henry Lambright, *Powering Apollo: James E. Webb of NASA* (Baltimore: Johns Hopkins, 1995), p. 210.

ACKNOWLEDGMENTS

313 **"The generation that came of age"** Arthur M. Schlesinger, Jr., interview in "NASA Audio News Feature: Apollo 11 Legacy on the Fifth Anniversary" (July 16, 1974).

ROBERT STONE is an Oscar- and Emmy-nominated documentary filmmaker. He and his work have been profiled in *The New York Times, The New Yorker,* and *Entertainment Weekly,* among many other publications.

robertstoneproductions.com
Twitter: @RobertStoneFilm

ALAN ANDRES spent three decades in trade book publishing and managed a small magazine enterprise. He served as a consulting producer and researcher on PBS's *Chasing the Moon.*

Twitter: @ChasingMoonBk

ABOUT THE TYPE

This book was set in Fairfield, the first typeface from the hand of the distinguished American artist and engraver Rudolph Ruzicka (1883–1978). Ruzicka was born in Bohemia (in the present-day Czech Republic) and came to America in 1894. He set up his own shop, devoted to wood engraving and printing, in New York in 1913 after a varied career working as a wood engraver, in photoengraving and banknote printing plants, and as an art director and freelance artist. He designed and illustrated many books, and was the creator of a considerable list of individual prints—wood engravings, line engravings on copper, and aquatints.